S0-ARK-137

PRINCIPLES OF TECHNOLOGY

UNIT 7

FORCE TRANSFORMERS

SECOND EDITION

Developed by the Center for Occupational Research and Development
in cooperation with the Agency for Instructional Technology and in
association with a consortium of state and provincial education agencies.

Library of Congress Cataloging-in-Publication Data

Principles of technology / developed by the Center for Occupational
    Research and Development in cooperation with the Agency for
    Instructional Technology and in association with a consortium of
    state and provincial education agencies. — 2nd ed.
        p.   cm.
    ISBN 1-55502-371-1 (set)
    1. Physics—Study and teaching (Secondary)   2. Mathematics—Study
and teaching (Secondary)   3. Technology—Study and teaching
(Secondary)   I. Center for Occupational Research and Development
(U.S.)   II. Agency for Instructional Technology
QC30.P75   1991
530—dc20   90-15001
                                                                CIP

ISBN 1-55502-378-9 (Unit 7, Force Transformers)

Printed in USA April 1997

Table of Contents

To the Student

More than ever before, technology is changing the way we live, the way we work, and the way we play. We read about the rapid advances of technology in lasers, telecommunications, and medicine. We sense the impact of technology on the changing workplace, where robots and automated manufacturing processes are becoming increasingly important. We see advancing technology in computerized offices and automobiles. And we stand back in awe when technology comes together in something as earth-shaking as a space vehicle. If we are to keep up with technology and control it, we need to understand it. *Principles of Technology* is designed to help you gain that understanding.

Principles of Technology is a course in applied science for those who plan to pursue careers as technicians or who just want to keep pace with the advances in technology. It blends an understanding of basic principles with practice in practical applications. It will give you a firm foundation for understanding the technology that surrounds you today—as well as the technology that's coming tomorrow.

Principles of Technology is made up of 14 units, each of which focuses on one of the important concepts that undergird modern technology—concepts such as force, work, rate, resistance, energy, or power. Each unit explains how that concept applies to mechanical, fluid, electrical, and thermal systems.

Each unit builds on earlier units, and together the units help you understand complex systems—like robots—in which mechanical, fluid, electrical, and thermal subsystems work together.

Principles of Technology is an opportunity to learn about modern technology—and the basic ideas that control and shape technology.

So learn well. Your future depends on it.

Leno S. Pedrotti

Preface to the Second Edition

Since its introduction in 1984, *Principles of Technology* (PT) has gained wide acceptance as a course in applied science (physics) for high school students and others. Today, thousands of creative teachers are using *Principles of Technology* to make scientific concepts understandable through hands-on learning in challenging laboratory settings. Across the country students are discovering that *Principles of Technology* provides them with a basic preparation for advancement to higher levels of education—and for a more successful entry into the world of work.

This new edition of *Principles of Technology* incorporates many of the helpful suggestions received from teachers who have implemented the course. First, the 90 hands-on laboratories—representing the core of the two-year course—have been modified and improved in response to this feedback. Second, a *Student Resource Book* has been developed as an aid to the student. This book contains a set of useful formulas and tables, a complete glossary, and the 18 remedial mathematics laboratories originally found in the appendices of Units 1 through 5 of the first edition. Finally, first edition errors in the printed text have been corrected, and unclear or inaccurate language has been improved or changed.

Despite these changes, the second edition of *Principles of Technology* remains essentially the same course of instruction as originally developed—*video, physics text for students and teachers, mathematics practice laboratories*, and *hands-on technology laboratories*. As mentioned above, several of the technology laboratories have undergone substantive changes—reflecting a concern for better results, improved equipment reliability, and increased safety. While the overall equipment needed for the PT laboratories has been affected by these changes, the changeover cost has been kept relatively minor. All established *Principles of Technology* equipment vendors have been advised of the changes and should be able to help schools make whatever laboratory modifications are necessary.

The course developers are grateful to the many experienced teachers of *Principles of Technology* who contributed significantly to the development of the second edition. A particular expression of thanks is extended to a special committee of teachers who met in Indiana in the Spring of 1989 to examine the priority and process of laboratory revisions. This dedicated group included: *David Crane*, North Carolina; *Chris Ennis*, Ohio; *Jim Everett*, Missouri; *Dick Jones*, Utah; *Al Kane*, Illinois; *Mark Kincaid*, Texas; *David Moore*, Georgia; *Art Reese*, Washington; *Keith Ross*, Colorado; *Della Schiebold*, Oregon; *George Taliadouros*, Massachusetts; and *Brad Thode*, Idaho.

To these and to the many others who provided us with the assistance needed to prepare a new edition, the course developers express their sincere appreciation. We hope that through our combined efforts, *Principles of Technology* will continue to serve the needs of those students who learn best by doing—who profit from those courses in science and mathematics that stress hands-on learning in the context of practical, world-of-work applications.

Leno S. Pedrotti
Fall 1990

Overview

UNIT OBJECTIVES

When you've finished the four subunits of Unit 7, "Force Transformers," you should be able to do the following:

1. *Describe force transformers in general. Describe force transformers in mechanical, fluid and electrical systems.*

2. *Explain why force transformers form a unifying principle in mechanical, fluid and electrical systems.*

3. *List examples of force transformers in mechanical, fluid and electrical systems.*

LEARNING PATH

1. *Read the "Overview" section. Give particular attention to the Unit Objectives.*

2. *View and discuss the video, "Overview: Force Transformers."*

3. *Participate in class discussions.*

MAIN IDEAS

- *Force transformers change—or transform—an input "force" to an output "force."*

- *In most cases, the output force delivered by a force transformer is larger than the input force. This provides the user with a <u>mechanical advantage</u>.*

- *IDEAL mechanical advantage is the "gain" a force transformer is said to have in the <u>design</u> stage. An ideal force transformer is one where NO friction—or resistance—is present.*

- *Actual mechanical advantage is the measured "gain" for a real force transformer. Actual mechanical advantage or "gain" is less than ideal mechanical advantage, because, in real life, friction/resistance reduces gain.*

- *Force transformers are used in mechanical, fluid and electrical systems. Basically, force transformers work the same way in each system. Therefore, if you know how a force transformer works in one system, you can understand how a force transformer works in another system.*

A Technician Talks...

I'm a technician in a plant that makes synthetic fiber products. Our products reflect the times. They range from a line of hand towels and diaper liners to protective jacket liners for computer discs.

For some time, our plant has been upgrading plant machinery and worker skills to meet stiff competition from overseas imports. Our union and management started a joint training program that's helping us learn to maintain our new machines. These machines are faster and more complex than our old machines. The new training program helps people like me—who have been out of school for awhile—update our skills.

It's interesting. We spend lots of time on the details of the new machines. But we're spending more time on the principles that tell us _why_ things work the way they do. We're learning how to troubleshoot and repair the new machines—even modify them in some cases.

As a maintenance technician, I hadn't really thought about the relationship between the different devices. Until now, devices either got the job done—or they didn't.

For example, all the levers, gears, pulleys, hoists and electrical transformers—as well as hydraulic pressure boosters in the plant—are different types of _force transformers_. And all of these mechanical devices have a common operating principle. They involve a coupling device, or transformer that works as a "helper."

For instance, a transformer takes an input from some source and changes it in some way. This produces an output that moves a load. The transformer is used to gain a mechanical _advantage_.

From our loading dock—with its ramp or inclined plane—to the hoist we use to lift large rolls of synthetic fiber, force transformers make our work easier. Small motors can operate big machines. That's because of the mechanical advantage we've gained in mechanical, fluid and electrical devices throughout the plant.

Since I know how force transformers work, troubleshooting is easier when machines malfunction. Now I know what purpose force transformers serve in machinery. So I can solve the problem sooner with fewer sidetracks. When troubleshooting is quick and correct, all that remains is "fixing the problem."

As our efficiency improves, our productivity increases. Therefore, our products become more competitive on the world market. That's good for our plant—and good for workers like me.

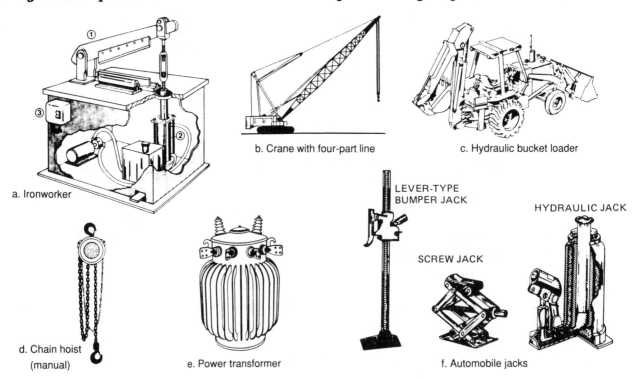

a. Ironworker

b. Crane with four-part line

c. Hydraulic bucket loader

d. Chain hoist (manual)

e. Power transformer

LEVER-TYPE BUMPER JACK

SCREW JACK

HYDRAULIC JACK

f. Automobile jacks

Fig. 7-1 Force transformer uses.

WHAT'S A FORCE TRANSFORMER?

The technician in the synthetic fiber products plant told you what force transformers *do.* But the technician told you very little about what they *are.*

Force transformers are simple machines or devices in mechanical, fluid and electrical energy systems that change input values of force, movement, or rate into different output values.

Force transformers are never 100% efficient. However—particularly in electrical systems—they sometimes approach that value.

Figure 7-1 shows common force transformer uses. In practice, a machine usually has many complex parts and movements. In fact, a machine may house more than one force transformer.

The ironworker of Figure 7-1a is a good example. It has three force transformers: (1) a mechanical lever, (2) a hydraulic cylinder and (3) a 240-volt to 110-volt electrical control transformer.

Look closely at the other examples in Figure 7-1. Try to see how many force transformers are included. At this point, you may see only a few. But when you finish this unit, look at Figure 7-1 again. You should be able to find more of the force transformers that are part of this machinery.

WHAT'S THE UNIFYING PRINCIPLE?

Force transformers have many uses. The unified approach you've learned to apply in *Principles of Technology* will also help you understand force transformers. Each energy system—mechanical, fluid and electrical—uses force transformers that work on the same physical principle. The unifying principle that links them is shown in Figure 7-2.

HOW DO FORCE TRANSFORMERS WORK IN MECHANICAL SYSTEMS?

Consider a mechanical system. Figure 7-1f shows different auto jacks. These mechanical force transformers are used by almost everyone. Figure 7-3 shows how the screw jack fits the unifying idea of Figure 7-2.

The *source* is the person who works the jack. The *input* is the force exerted by the person's arm. The *coupling device* is the screw jack. The *load* is the car. The *output* is the force applied to raise one side of the car.

The screw jack magnifies (boosts) the input force. It provides an output force strong enough to lift the car. A screw jack is a type of *inclined plane* or ramp, a common mechanical force transformer.

The schematic in Figure 7-3 also applies to a bumper jack. However, a bumper jack uses a *lever action.* Levers and inclined planes are two of the most basic mechanical force transformers. Therefore, the lever and the inclined plane (screw) are two mechanical transformers you'll study in Unit 7.

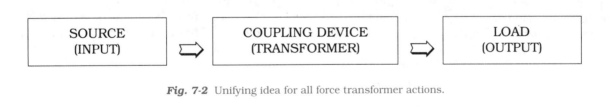

Fig. 7-2 Unifying idea for all force transformer actions.

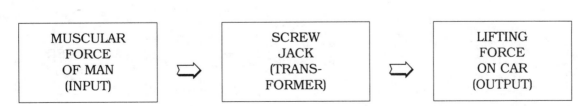

Fig. 7-3 Typical schematic for a mechanical force transformer.

HOW DO FORCE TRANSFORMERS WORK IN FLUID SYSTEMS?

A fluid "force" transformer is a pressure booster—or pressure intensifier. In pressure boosters, input force is created when a low-pressure fluid pushes on the surface of a large piston in the input chamber. This piston mechanically moves the same input force to a small output piston surface. The same force, when applied to a much smaller area, creates a high pressure in the output chamber fluid. (See Figure 7-4.)

| LOW-PRESSURE FLUID SOURCE (INPUT) | ⇨ | FLUID PRESSURE INTENSIFIER (TRANSFORMER) | ⇨ | HIGH-PRESSURE FLUID TOOL (OUTPUT) |

Fig. 7-4 Fluid force transformers follow the unifying idea.

To simplify how a pressure intensifier operates, let's compare flat shoes and high-heeled shoes. When your body's weight is supported by a larger area (as it is when you're wearing flat-heeled shoes) the force per unit area (pressure) on your heels isn't too large. By contrast, when your body weight is supported by a smaller area (as when you're wearing high-heeled shoes) the force per unit area is large. Therefore, the pressure on the heels of the shoes increases.

So the unifying idea shown in Figure 7-2 is true for the pressure booster that acts as a fluid transformer. That's because pressure is a "forcelike quantity." Input pressure is amplified (made bigger) by the "force" transformer. This gives a higher output pressure.

It's tempting to think that a hydraulic jack used to raise heavy loads is also a *fluid* "force" transformer. However, a hydraulic jack is—in fact—a *mechanical* force transformer. This is true, even though the hydraulic jack uses fluid to transmit and apply the force. A hydraulic jack is a mechanical force transformer because it transforms (increases) mechanical *force*—not *pressure*.

HOW DO FORCE TRANSFORMERS WORK IN ELECTRICAL SYSTEMS?

Now let's consider electrical systems. Voltage difference is the "forcelike quantity."

Many portable devices (calculators, radios, cassette players, etc.) run on 6-volt batteries. To save battery drainage, rechargeable batteries and a "step-down" electrical transformer are used. When a 110-volt line voltage is available, it's stepped down to 6 volts by an electrical "force" transformer.

The basic idea of force transformers in electrical systems is shown in Figure 7-5. Note how this idea follows the general form shown in Figure 7-2.

Electrical force transformers are very useful. They come in many sizes and ratings. These sizes and ratings are designed to match the intended use for each transformer.

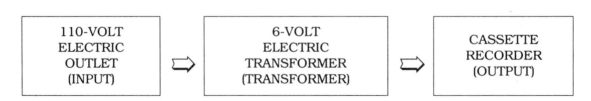

| 110-VOLT ELECTRIC OUTLET (INPUT) | ⇨ | 6-VOLT ELECTRIC TRANSFORMER (TRANSFORMER) | ⇨ | CASSETTE RECORDER (OUTPUT) |

Fig. 7-5 Electrical force transformers follow the unifying idea.

In all the energy systems you've studied, energy is *produced* by input work done at the source. Energy is *consumed* as output work done on the load. From the unifying equation for work, you know that the "force" and "displacement" involved in the process of doing work relate as follows:

Work = Forcelike Quantity × Displacementlike Quantity   ***Equation 1***

A force transformer, or coupling device, moves energy from the source to the load. The forcelike quantities and displacementlike quantities at the input end and output end have different magnitudes.

However, in the ideal case of no resistance the **product** of these magnitudes is the same. That's because

$$F_i \times D_i = F_o \times D_o$$

where: F_i = input force
D_i = input displacement
F_o = output force
D_o = output displacement

Suppose, for a certain transformer, that the input force (F_i) is 5 lb, and the input displacement (D_i) is 10 ft. At the same time, the output force (F_o) is 25 lb and the output displacement (D_o) is 2 ft. By substituting in the values given, you get:

$$F_i \times D_i = F_o \times D_o$$
$$5 \text{ lb} \times 10 \text{ ft} = 25 \text{ lb} \times 2 \text{ ft}$$
$$50 \text{ ft·lb} = 50 \text{ ft·lb}$$

Sure enough, "input work" equals "output work." However, note that the output force of 25 pounds is five times greater than the input force. This is offset, exactly, by the difference in displacements. The output displacement is only one-fifth of the input displacement.

In real life, **input work** doesn't quite equal **output work,** because some input work—or energy—is lost in the transformer. This loss is due to friction or other kinds of resistance. The "lost" energy is dissipated as heat. Keeping heat losses low is one of the chief goals of machine designers, engineers and technicians.

Equation 2 shows how to find force transformer *efficiency*:

$$\text{Eff} = \frac{\text{Work Out}}{\text{Work In}} \times 100\% \qquad\qquad \textbf{\textit{Equation 2}}$$

where: Work Out = useful work at output
Work In = required work at input
100 = multiplier to obtain %

When useful work out equals work in, no energy is lost in the coupling device. This is called an ***"ideal transformer."*** An ideal transformer is always assumed in the design stage of building a force transformer. On the drawing board, the efficiency is always assumed to be 100%. In practice, some energy is always lost in force transformers. So actual efficiency is less than 100%.

Mother Nature never gives up something for nothing. For instance, a bumper jack (a mechanical force transformer) may amplify an input force into a larger output force. However, it does so at the expense of displacement.

For example, to use a jack to increase the force, your hand may have to move up and down on the end of the jack handle a distance of 12 to 14 inches each stroke. The jack may raise the car only about one inch per stroke. Even so, total energy is conserved. So remember, increased force at the output is offset by increased movement at the input.

Figure 7-6 shows other force transformers technicians use.

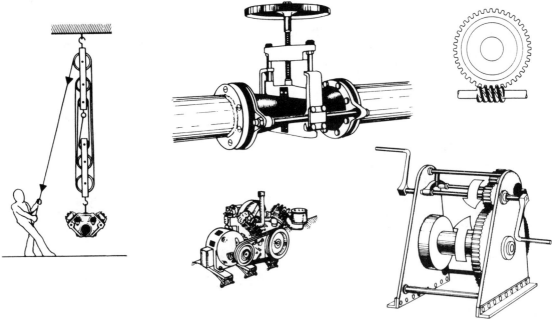

Fig. 7-6 Other force transformers.

In the following subunits, you'll learn about force transformers in mechanical, fluid and electrical systems. Study the action of each force transformer as follows:

- First, try to pick out **what** is being transformed or changed (force, torque, speed, pressure, etc.).
- Next, try to understand **how** it is being transformed (increased, decreased, etc.).
- Then identify **what** is being sacrificed (the offset) to make the change.

Force Transformers in Linear Mechanical Systems

SUBUNIT OBJECTIVES

When you've finished reading this subunit and viewing the video, "Force Transformers in Linear Mechanical Systems," you should be able to do the following:

1. Explain the relationship between Input Work and Output Work for linear force transformers.

2. Use the equation, Work In = Work Out, to find an unknown force or displacement when using a particular force transformer.

3. Explain the difference between ideal mechanical advantage (IMA) and actual mechanical advantage (AMA).

4. Find the ideal mechanical advantage and actual mechanical advantage of a force transformer.

5. Determine the efficiency of a force transformer.

6. Explain the difference between three classes of levers.

7. List or identify different kinds of force transformers for linear mechanical systems.

8. Identify workplace applications where technicians use force transformers.

9. Measure the mechanical advantage of a linear force transformer.

LEARNING PATH

1. Read this subunit, "Force Transformers in Linear Mechanical Systems." Give particular attention to the Subunit Objectives.

2. View and discuss the video, "Force Transformers in Linear Mechanical Systems."

3. Participate in class discussions.

4. Watch a demonstration about linear mechanical force transformers.

5. Complete the Student Exercises.

MAIN IDEAS

- *Linear mechanical force transformers involve a force and displacement at the <u>input</u> end—and a force and displacement at the <u>output</u> end.*

- *Mechanical force transformers usually amplify force at the expense of displacement. When a force transformer amplifies force, it develops a mechanical advantage.*

- *Common linear force transformers include the pulley, the lever and the inclined plane.*

- *Work input equals work output in an ideal—or 100% efficient—force transformer.*

- *Mechanical advantage is expressed as ideal (IMA) or actual (AMA) by the following equations: $IMA = D_i/D_o$ and $AMA = F_o/F_i$.*

WHAT'S AN IDEAL TRANSFORMER?

An **ideal transformer** is one that's 100% efficient. For *ideal* transformers, **work out** always **equals work in.** But in real-world uses, force transformers are never 100% efficient. That's because resistance is present—and **work out** is always **less than work in.**

A transformer's true efficiency always depends on how much resistance and friction are present during the actual operation. Nevertheless, by assuming that ideal, 100%-efficient transformers exist, you can simplify the analysis and determine their best possible performance.

So, when measurements are made on a real force transformer, don't be surprised when the efficiency comes out less than 100%.

LET'S LOOK AT TRANSFORMERS IN LINEAR MECHANICAL SYSTEMS

In the *Overview,* you learned the unified concept of a force transformer. You saw a schematic of the relationship between the *input,* the *coupling device,* and the *output* (Figure 7-2). Let's examine that schematic in terms of a linear mechanical energy system. From Figure 7-2:

SOURCE $\Longrightarrow$ COUPLING DEVICE $\Longrightarrow$ LOAD
(INPUT)   (TRANSFORMER)   (OUTPUT)

Consider a linear mechanical system. The input to a force transformer is the work done at the input. This is linear work. Linear work equals applied input force F_i times the input displacement D_i. This relationship can be written as follows:

$$\text{Work In} = F_i \times D_i$$

A force transformer changes input work to output work that moves the load. For an ideal transformer, input work equals output work. That means that $F_i \times D_i$ equals $F_o \times D_o$. But even though the products are equal, F_i doesn't equal F_o and D_i doesn't equal D_o.

The force transformer increases the value of F. So F_o is **greater** than F_i. At the same time, it decreases the value of D. So D_o is **less** than D_i. It's similar to the relation $2 \times 5 = 5 \times 2$. Both products equal 10. But the first number on the left (2) increases to 5 on the right. At the same time, the second number on the left (5) decreases to 2 on the right.

For an **ideal** transformer in a linear mechanical system, you can sum up the basic work equation as follows.

$$\text{Work Input} = \text{Work Output}$$

For ideal linear mechanical systems, this becomes

$$F_i \times D_i = F_o \times D_o \qquad \textbf{Equation 3}$$

where: F_i = input force
$\quad\quad\quad D_i$ = input displacement
$\quad\quad\quad F_o$ = output force on load
$\quad\quad\quad D_o$ = output load displacement

There are many linear force transformers in which inputs and outputs closely satisfy Equation 3. Two common mechanical linear force transformers are the lever and the pulley (block and tackle). Let's start our discussion with pulley systems.

A pulley system (sometimes called a "block and tackle") is a linear force transformer that's made up of two or more pulleys. To understand how pulley systems work, let's study Figure 7-7.

In Figure 7-7a, a weight is lifted by pulling on a cord wrapped around a **fixed** pulley. If the fixed pulley is frictionless, the force (F) required to lift the weight is equal to the weight. Here, the purpose of this pulley is only to **change direction** of the pulling force.

Since $F_i = F_o$, force isn't changed, and there is no force advantage. Therefore this arrangement isn't a force transformer. In addition, if one foot of cord is pulled through the pulley, the weight rises one foot ($D_i = D_o$). So displacement isn't changed, either.

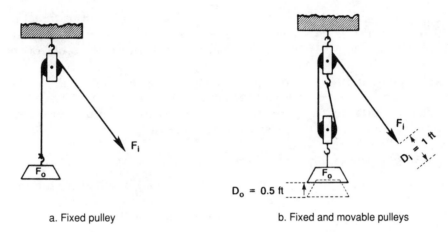

a. Fixed pulley   b. Fixed and movable pulleys

Fig. 7-7 Pulley systems

The multiple-pulley arrangement in Figure 7-7b **is** a force transformer. It's often called a "block and tackle." Study this machine. Why is $D_o = \frac{1}{2} D_i$?

Moving the rope a distance D_i with an input force F_i results in movement of the weight a distance D_o by the output force F_o. F_i times D_i is the input work. F_o times D_o is the output work. Since we're dealing with **ideal** force transformers, **input work equals output work.** This is summed up in Equation 3.

<u>Ideal Force Transformer</u>

Work In = Work Out   ***Equation 3***
                                 (given previously)

$$F_i \times D_i = F_o \times D_o$$

where: F_i = the input force at the source of the pulley system
         D_i = distance moved by input force
         F_o = output force at the load of the pulley system
         D_o = distance moved by output force or weight

Let's use Equation 3 and Figure 7-7b to see how the input and output forces compare. You also can see how the input and output displacements compare.

Dividing each side of Equation 3 by D_i gives:

$$\frac{F_i \times \cancel{D_i}}{\cancel{D_i}} = \frac{F_o \times D_o}{D_i}$$

Canceling the D_i terms on the left and regrouping the equation yields:

$$F_i = \left[\frac{D_o}{D_i}\right] \times F_o$$

As shown in Figure 7-7b, $D_o = \frac{1}{2}$ foot, when $D_i = 1$ foot. What should the input force (F_i) be if the weight (output force F_o) is 100 pounds? Let's substitute in the equation to find out.

$$F_i = \left(\frac{D_o}{D_i}\right) \times F_o$$

where: $D_o = 0.5$ ft
$D_i = 1$ ft
$F_o = 100$ lb

$$F_i = \left(\frac{0.5 \text{ ft}}{1 \text{ ft}}\right) \times 100 \text{ lb}$$

$$F_i = 50 \text{ lb}$$

Thus, it takes an input force of 50 pounds to lift a load (output force) of 100 pounds. The block and tackle is clearly a force transformer. This is because a force of 50 pounds on the input is transformed (increased) into a force of 100 pounds on the output.

As we've said, nature doesn't give up something for nothing. We've managed to magnify—or amplify force. The output force is TWICE the input force. But input force must act through a displacement two times as great as the output force. So the force transformer gains force at the expense of distance moved.

WHAT'S MECHANICAL ADVANTAGE?

A load twice as large as the applied input force can be lifted by the pulley system in Figure 7-7b. This means the pulley system "has a *mechanical advantage* of 2." For an *ideal* force transformer, **ideal mechanical advantage** (IMA) is given by Equation 4.

$$IMA = \frac{D_i}{D_o}$$ ***Equation 4***

where: IMA = ideal mechanical advantage
D_i = distance moved by input force
D_o = distance moved by output force or weight

Equation 4 says that the **ideal** mechanical advantage equals the ratio of distances moved (D_i/D_o). If the transformer is 100% efficient, so that Work In equals Work Out, you can obtain another ratio that's equal to D_i/D_o. Here's how.

$$\text{Work In} = \text{Work Out} \qquad \text{(for ideal transformer only)}$$
$$F_i \times D_i = F_o \times D_o$$

$$\frac{F_i \times D_i}{F_i D_o} = \frac{F_o \times D_o}{F_i D_o} \qquad \text{(Divide each side by } F_i D_o.)$$

Therefore, $$\frac{D_i}{D_o} = \frac{F_o}{F_i}$$

The force ratio F_o/F_i is equal to the displacement ratio D_i/D_o and the IMA only when the force transformer is "ideal" or 100% efficient. When resistance or friction is present, the force ratio F_o/F_i is **always less than** the displacement ratio D_i/D_o and the IMA.

If you use the ratio D_i/D_o, as done in Equation 4, you find the **ideal mechanical advantage.** If you use the ratio F_o/F_i, as shown in Equation 5, you find the **actual mechanical advantage.**

$$AMA = \frac{F_o}{F_i}$$ ***Equation 5***

where: AMA = actual mechanical advantage
F_o = output force
F_i = input force

Remember, IMA and AMA are equal *only* when no *resistance* exists in a force transformer. IMA depends only on design. AMA depends on actual performance.

Example 7-A shows how mechanical advantage is found in a pulley system (block and tackle) on a wrecker.

Example 7-A: *Mechanical Advantage of a Block and Tackle on a Wrecker* ————————

Given: Bill Johnson operates a wrecker service. The winch cable on his wrecker is "rated" at 10 tons. (Because there's a 200% safety factor, the cable can actually pull 20 tons.) To "winch" a truck back onto the road, Bill decides to rig the cable into a block and tackle (as shown in the diagram). The winch winds in 300 feet of cable while applying a force of 10 tons. At the same time, the truck is pulled a distance of 150 feet back onto the road.

Find: a. The work done by the winch.

b. The force applied by the block and tackle to the load (that is, to the truck).

c. The work done on the truck.

d. The mechanical advantage of the block and tackle. Assume 100% efficiency.

Solution: a. If you assume 100% efficiency (ideal transformer), then Work In = Work Done on the block and tackle by the winch.

Work In = $F_i \times D_i$

W_i = 20,000 lb × 300 ft   (10 tons = 20,000 pounds)

W_i = $(2 \times 10^4)(3 \times 10^2)$ (ft·lb)

W_i = 6×10^6 ft·lb

b. To find F_o, start with Work In = Work Out, and solve for F_o.

$F_i \times D_i$ = $F_o \times D_o$

Solve for F_o.

$$F_o = \frac{F_i D_i}{D_o}$$ (F_o is the force applied by block and tackle to the load.)

$$F_o = \frac{W_i}{D_o}$$ where $W_i = 6 \times 10^6$ ft·lb and $D_o = 150$ ft

$$F_o = \frac{6 \times 10^6 \text{ ft·lb}}{150 \text{ ft}}$$

$$F_o = \left(\frac{6 \times 10^6}{1.5 \times 10^2} \right) \text{lb} = 4 \times 10^4 \text{ lb} = 40,000 \text{ lb}$$

c. Work output of block and tackle = Work done on truck

Work Out = $F_o D_o$

W_o = 40,000 lb × 150 ft

W_o = $(4 \times 10^4)(1.5 \times 10^2)$ (ft·lb)   ($W_o = W_i$ for 100% efficiency)

W_o = 6×10^6 ft·lb

d. To find the ideal mechanical advantage (IMA)

$$\text{IMA} = \frac{D_i}{D_o} = \frac{300 \text{ ft}}{150 \text{ ft}} = 2$$

Thus, the ideal mechanical advantage is 2.

Note: Where 100% efficiency of the transformer is assumed, IMA = AMA and either the force ratio (F_o/F_i) or the distance ratio (D_i/D_o) yields the same result. To show that this is true, use Equation 5 and the force ratio as follows:

$$\text{AMA} = \frac{F_o}{F_i} = \frac{4 \times 10^4 \text{ lb}}{2 \times 10^4 \text{ lb}} = 2$$

Thus, the actual mechanical advantage is also 2.

HOW DO YOU FIND THE IMA OF A BLOCK-AND-TACKLE TRANSFORMER BY "COUNTING SUPPORTING ROPES"?

Knowing the ideal mechanical advantage of a force transformer is important. The ideal mechanical advantage can tell you, for example, the heaviest possible load you can lift with a given force transformer.

To find the ideal mechanical advantage of any block and tackle, simply count the number of ropes or lines that **support** the load (w). See Figure 7-8. A single pulley can be used to gain mechanical advantage only when it's rigged as a movable pulley, as shown in Figure 7-8b; *not* as in Figure 7-8a.

Note that the actual mechanical advantage will be less. That's because the block and tackle has some frictional resistance in the pulley movement that must be overcome. So the real F_o/F_i is less than the ideal F_o/F_i.

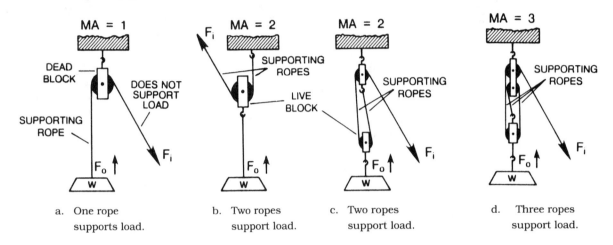

a. One rope supports load.

b. Two ropes support load.

c. Two ropes support load.

d. Three ropes support load.

Fig. 7-8 Ideal mechanical advantage.

HOW DO YOU FIND THE EFFICIENCY OF A FORCE TRANSFORMER?

You can use several equations to find the efficiency of a force transformer. Two of the more useful forms can be seen in Equations 2 and 6. Either equation gives the same result.

$$\text{Eff} = \frac{\text{Work Out}}{\text{Work In}} \times 100\% \qquad \textbf{Equation 2}$$
(given previously)

where: Work Out = $F_o \times D_o$ at output end of force transformer
Work In = $F_i \times D_i$ at input end of force transformer

$$\text{Eff} = \frac{\text{AMA}}{\text{IMA}} \times 100\% \qquad \textbf{Equation 6}$$

where: AMA = actual mechanical advantage = F_o/F_i
IMA = ideal mechanical advantage = D_i/D_o

Example 7-B shows how these equations are used to find the efficiency of a block and tackle.

Example 7-B: *Mechanical Advantage and Efficiency of the Block and Tackle* ─────

Given: In Example 7-A, **where no friction was assumed to be present,** an applied input force of 10 tons (20,000 lb) was changed by the block and tackle into an output force of 40,000 lb. However, **because friction is present,** it really takes 11 tons (22,000 lb) of applied force to produce 40,000 at the output end.

Find: a. The actual mechanical advantage of the block and tackle.

b. The efficiency of the block and tackle.

Solution: Remember, F_o/F_i always gives the *actual* mechanical advantage.

a. $\text{AMA} = \frac{F_o}{F_i} = \frac{40,000 \text{ lb}}{22,000 \text{ lb}} = 1.818$

b. Using Equation 2,

$$Eff = \frac{AMA}{IMA} \times 100\%$$

where: AMA = 1.818 (from Part a)

IMA = 2 (from Example 7a)

$$Eff = \frac{1.818}{2} \times 100\%$$

$$Eff = 90.9\%$$

Using Equation 6,

$$Eff = \frac{W_o}{W_i} \times 100\%$$

where: Work Out = W_o = 6×10^6 ft·lb
(from Example 7-A)

Work In = W_i = $F_i \times D_i$

(D_i = 300 ft, from Example 7-A.)

Find W_i as follows:

$$W_i = 22,000 \text{ lb} \times 300 \text{ ft}$$
$$W_i = 6.6 \times 10^6 \text{ ft·lb}$$

So,

$$Eff = \frac{6 \times 10^6 \text{ ft·lb}}{6.6 \times 10^6 \text{ ft·lb}} \times 100\%$$

$$Eff = 90.9\%$$

Thus, both equations for efficiency give the same answers—as they should!

THE LEVER IS A LINEAR FORCE TRANSFORMER.

All levers are used for the same purpose—to gain a mechanical advantage. The lever is a rigid bar that may be pivoted at some point. All levers belong to one of three classes.

Two things determine the lever class. First, check the location of the pivot point along the bar. Second, check where the input force and output force are located with respect to the pivot point.

Figure 7-9 shows the three classes of levers. They're referred to as **first-class, second-class** and **third-class** levers. Note carefully the relative position of the **effort** (F_i), the **load** (F_o) and the **pivot**. Note how they are different for each class of lever.

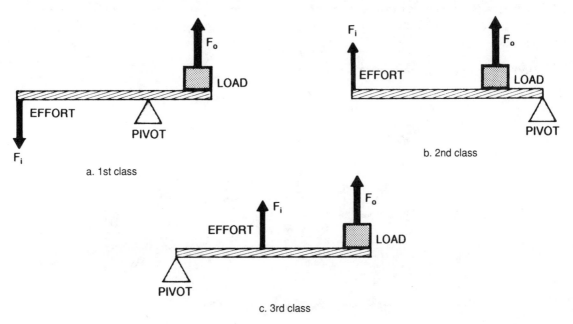

Fig. 7-9 The three classes of levers.

WHAT'S A FIRST-CLASS LEVER?

In a first-class lever, the input force F_i and output force F_o are always on opposite sides of the pivot. One common first-class lever is a seesaw (or teeter-totter). In the drawing, Dora's weight is the input force. Her weight equals the combined weight of Billy and the robot—the output force.

As shown, Dora is as far from the pivot on one side as Billy and his robot are on the other side. The lever arms are equal. Therefore, the seesaw is in equilibrium. Clockwise and counterclockwise torques are *balanced.*

When Dora drops a distance of one foot on one end, Billy and his robot rise exactly one foot on the other end. This means that Work In and Work Out are equal, or that $F_i \times D_i = F_o \times D_o$.

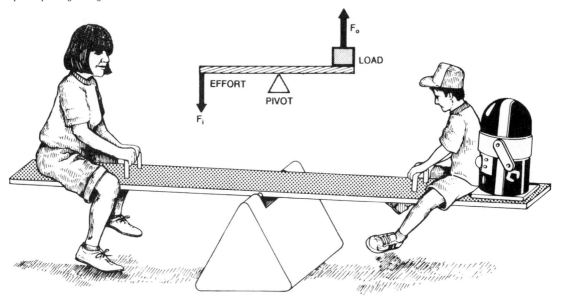

Let's look at these ideas in the two examples that follow.

Example 7-C: *Truck Scale—A First-class Lever*

Given: A truck rolls onto the platform of a truck-weighing scale. The truck weight (w) acts on a 0.5-ft lever arm about the pivot point. A weight of 1200 lb, located 20 ft from the pivot, balances the truck scale.

Find: The weight of the truck on the scale.

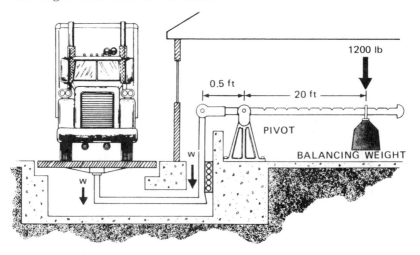

Solution: Since the scale is balanced, the torques are balanced. This means that the clockwise (cw) torque about the pivot is equal to the counterclockwise (ccw) torque about the pivot.

$$\text{ccw Torque} = \text{cw Torque}$$

The "cw" torque is 1200 lb × 20 ft. (The lever arm of "1200 lb" is 20 ft.)

The "ccw" torque is w × 0.5 ft. (The lever arm of "w" is 0.5 ft.)

Equating torques, w × 0.5 ft = 1200 lb × 20 ft.

Solve for w.

$$w = \frac{1200 \text{ lb} \times 20 \text{ ft}}{0.5 \text{ ft}}$$

$$w = \left(\frac{1200 \times 20}{0.5}\right)\left(\frac{\text{lb·ft}}{\text{ft}}\right)$$

$$w = 48,000 \text{ lb}$$ (Notice that $^w/_{1200} = {}^{20}/_{0.5} = 40$.)

The truck scale has "weighed" the truck and found it to weigh 48,000 lb. As in the example of the "seesaw," balanced torques create a condition of equilibrium (balance).

Example 7-C shows a lever with an IMA of $^{48,000}/_{1200} = 40$. That's exactly the ratio of the lever arms: $^{20}/_{0.5} = 40$. The IMA of a lever can always be found in this way: IMA = Input lever arm/Output lever arm = L_i/L_o.

Now let's look more closely at the seesaw (first-class lever) to study the relationship of Work In = Work Out. Example 7-D will help you understand this relationship better.

Example 7-D: *Seesaw as a Force Transformer (First-class Lever)* ————————

Given: A 100-pound child 4 feet from the pivot of a seesaw balances an 80-pound child 5 feet from the pivot.

Find: How far the 100-pound child moves upward when the 80-pound child moves downward 2 feet.

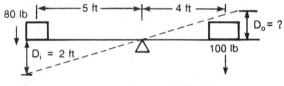

Solution: Let the input side of the force transformer be the end where the 80-lb child sits. Then,

Note: Let children be represented by weights.

F_i = 80 lb. The output side is where the 100-lb child is, so F_o = 100 lb. You're told that the 80-lb force (F_i) moves 2 ft. So D_i = 2 ft. You need to know the distance (D_o) the 100-lb weight moves.

The seesaw has very little friction. So, you can safely assume that it's 100% efficient. Therefore, Work Out = Work In. That's the key to this solution.

Work Out = Work In

$$F_o \times D_o = F_i \times D_i$$

Solve this equation for D_o, the unknown distance.

$$\frac{\cancel{F_o} \times D_o}{\cancel{F_o}} = \frac{F_i \times D_i}{F_o}$$

$$D_o = \frac{F_i \times D_i}{F_o}$$

Substitute in values for F_i, D_i and F_o where F_i = 80 lb, D_i = 2 ft, and F_o = 100 lb.

$$D_o = \frac{80 \text{ lb} \times 2 \text{ ft}}{100 \text{ lb}} = \left(\frac{80 \times 2}{100}\right)\left(\frac{\cancel{lb}\cdot\text{ft}}{\cancel{lb}}\right)$$

$$D_o = 1.6 \text{ ft}$$

Thus, the output force (100 lb) moves a distance of 1.6 ft while the input force (80 lb) moves a distance of 2 ft. Again, you can see the trade-off. The seesaw *transforms* an 80-lb force into a 100-lb force. But the 80-lb force travels 2 ft to lift the 100-lb force (weight) a distance of 1.6 ft.

A seesaw is one example of a first-class lever. It's playground equipment. It's not a useful force transformer. However, a 5-foot wrecking bar is a first-class lever that's a useful force transformer. It can be used with a pivot (such as a rock, a log or other object) placed under it. A wrecking bar also has a built-in pivot point.

The wrecking bar has a large mechanical advantage, due to the length of lever arm L_o compared to lever arm L_i. Figure 7-10 shows this. By the way, another common technical term for pivot point is **fulcrum.** "Pivot" and "fulcrum" mean the same thing.

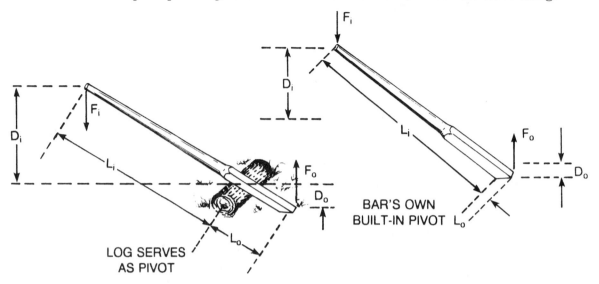

Fig. 7-10 Mechanical linear force transformer (5-ft wrecking bar).

When friction is small, the efficiency of a 5-ft wrecking bar nears 100%. On a level railroad track, a 5-ft bar often is used to move several loaded rail cars. The pivot point of the bar is put on the track behind the wheel. A small downward force on the long end of the lever (Figure 7-10b) produces a large force (F_o) on the wheel. This causes the wheel to turn and move the rail cars.

First-class levers always involve opposing torques. They come in many sizes and shapes. But first-class levers always involve forces applied at some distance on opposite sides of a pivot point.

WHAT ARE SECOND-CLASS LEVERS?

For a second-class lever, the effort F_i and the load F_o are on the same side of the pivot and the effort is farther away than the load. Figure 7-9b shows that the input lever arm L_i is the full length of the bar. The output lever arm L_o is the distance from the pivot back along the bar to where the output force F_o—or load—is applied.

Figure 7-11 shows a wheelbarrow. A wheelbarrow is a force transformer—a second-class lever. It's also a useful machine. The pivot point on the wheelbarrow is at the end, where the wheel touches the ground. The effort or input force F_i is at the opposite end.

The load (or output, labeled F_o) is between the pivot and input force. The lever arms are both measured from the pivot. Since L_i is longer than L_o, the wheelbarrow provides a mechanical advantage. That means that one can lift a heavy load (F_o) with a smaller force (F_i).

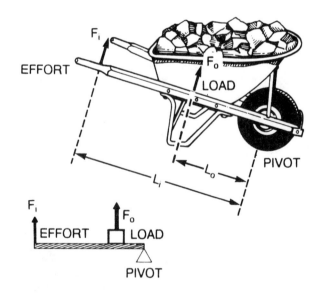

Fig. 7-11 A wheelbarrow as a second-class lever.

WHAT ARE THIRD-CLASS LEVERS?

For a third-class lever, the effort and load are again on the same side of the pivot. But now, in contrast with the second-class lever, the load is farther away than the effort. (To see that clearly, compare Figures 7-9b and 7-9c.) In a third-class lever, the load distance from the pivot (L_o) is greater than the input force distance (L_i) from the pivot. The mechanical advantage $L_i/L_o < 1$. As a result, when the input force is applied over a small displacement (D_i) the output force must act through a greater displacement (D_o).

A third-class lever increases **displacement** rather than **force.** In this case, the output force is smaller than the input force. (See Figure 7-12a.)

Figure 7-12b shows the use of a third-class lever. It shows the human arm. The muscle (F_i) is attached a distance (L_i) from the elbow (pivot). The hand is a distance (L_o) from the elbow. The force (F_o) that can be exerted by the hand is small compared to the input force (F_i) of the muscle. But the hand travels through a much greater distance (D_o) than the distance (D_i) through which the muscle contracts. The result is a small muscle movement that results in great speed at the hand.

Third-class levers often are used on industrial machines where an adjustable output movement with little input movement is needed. Machines that change linear feed rates of material being processed are common examples. The adjustable ratchet pawl (shown in Figure 7-12c) is a device that represents a third-class lever.

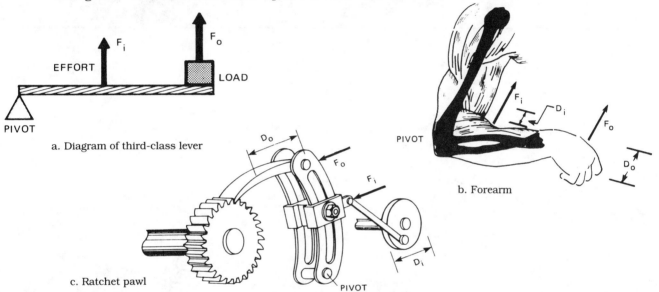

a. Diagram of third-class lever

b. Forearm

c. Ratchet pawl

Fig. 7-12 Third-class levers.

WHAT ARE SOME OTHER LINEAR FORCE TRANSFORMERS?

The pulley and the lever are two common linear force transformers. Another force transformer with many uses is the inclined plane. Common uses of the inclined plane include the loading ramp, the wedge (a double inclined plane) and the ordinary screw. In the screw, an inclined plane is "wrapped" around a shaft. It acts like a rotational force transformer. Let's discuss the use of the inclined plane as a ramp. Suppose a crate of weight w is to be loaded onto a truck. The crate can be loaded by lifting the crate from the ground vertically to the level of the truck bed. The crate can also be loaded by pushing it up a ramp (or inclined plane) from the ground to the truck bed.

Figure 7-13 shows the loading problem. If the inclined plane is frictionless, it takes the same amount of *work* to lift the crate straight up onto the truck bed as it does to slide up the ramp. However, the *force* needed to slide the box up the ramp over a longer distance is less than the force needed to lift it straight up through a smaller vertical height.

That's the advantage of the inclined plane (ramp). You can move something "up" with less force.

Suppose that the inclined plane (ramp) is frictionless. If that's the case, the ideal mechanical advantage (IMA) and the actual mechanical advantage (AMA) are equal. But most ramps have friction. So the actual mechanical advantage (F_o/F_i) will be less than the ideal mechanical advantage (D_i/D_o).

The ratio of the rise to the incline (D_o/D_i) is often called the **"grade"** of the incline. Thus, the ideal mechanical advantage (IMA) is easily remembered as the "reciprocal of the grade of the incline." Example 7-E shows how an inclined plane gives us a force advantage. Example 7-F shows how the wedge operates as an inclined plane.

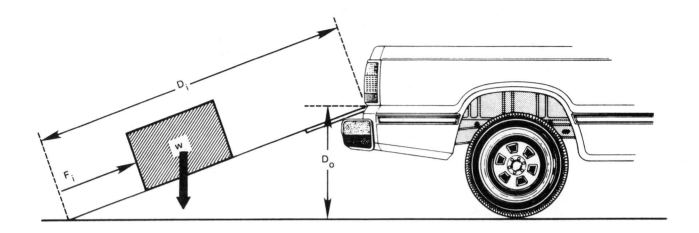

Fig. 7-13 An inclined plane or ramp.

Example 7-E: *Mechanical Advantage of an Inclined Plane*

Given: A 300-lb safe is pushed up an incline 15 ft long and 5 ft high on ***frictionless*** rollers.

Find:
    a. The grade of the incline.
    b. The ideal mechanical advantage of the incline.
    c. The force required to roll the safe up the incline.

Solution:

a. $\text{Grade} = \dfrac{\text{Incline Rise}}{\text{Incline Length}} = \dfrac{D_o}{D_i}$

$\text{Grade} = \dfrac{D_o}{D_i} = \dfrac{5 \text{ ft}}{15 \text{ ft}}$

$\text{Grade} = \frac{1}{3}$

b. The ideal mechanical advantage **is always** equal to $\dfrac{1}{\text{grade}}$. Therefore,

$\text{IMA} = \dfrac{1}{\text{grade}} = \dfrac{1}{\frac{1}{3}} = 3$

c. Since movement is frictionless, Work In = Work Out.

$F_i \times D_i = F_o \times D_o$

Solve for F_i, where $F_o = w = 300$ lb, $D_o = 5$ ft, and $D_i = 15$ ft.

$F_i = \dfrac{F_o \times D_o}{D_i}$

$F_i = \dfrac{300 \text{ lb} \times 5 \text{ ft}}{15 \text{ ft}} = \left(\dfrac{300 \times 5}{15}\right)\left(\dfrac{\text{lb} \cdot \text{ft}}{\text{ft}}\right)$

$F_i = 100$ lb   (This follows also from IMA = F_o/F_i.)

The force required to move the 300-lb safe is only 100 lb.

Example 7-F: *The Wedge as an Inclined Plane*

Given: A wedge used to split logs.

Find: Why the wedge makes the job easier.

Solution: The wedge is one use, or application, of the inclined plane. Notice that the wedge is just two inclined planes joined together, back-to-back. Its ideal mechanical advantage allows you to use a driving force (F_i) that's **smaller** than the sideways force (F_o) required to split the log. The ideal and actual mechanical advantages are as follows:

$$IMA = \frac{2\,h}{b}$$

$$AMA = \frac{2\,F_o}{F_i}$$

The meanings of "h" and "b" are shown in the diagram.

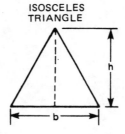

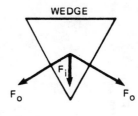

Question:

Do you think that the ideal mechanical advantage and the actual mechanical advantage of a wedge are about equal? Why?

Student Exercises

The following exercises review the main ideas and definitions presented in this subunit, "Force Transformers in Linear Mechanical Systems." Complete each question.

1. A force transformer has a source (or input), a coupling device or transformer and ____. (Complete the sentence.)

2. If a force transformer is 100% efficient, then "work in" ____ (is, isn't) equal to "work out."

3. The work input to a pulley force transformer is 100 ft·lb. The load moves 6 inches (0.5 ft) when the rope is pulled through 2 ft. What's the output force applied by the transformer?

4. In a **frictionless** block and tackle, a 50-lb input force raises a 300-lb weight 2 ft off the ground.
   a. How far was the rope pulled by the input force?
   b. What is the ideal mechanical advantage of the block and tackle?

5. In Problem 4, because the block and tackle actually did have some friction, it took 60 pounds of input force to raise the 300-pound weight 2 feet off the ground. What is the actual mechanical advantage of the block and tackle?

6. Using the ideal mechanical advantage from Problem 4b and the actual mechanical advantage from Problem 5, find the percent efficiency of the block and tackle.

7. Describe the three classes of levers by drawing a diagram of each. Label the three diagrams correctly. Also, add a statement about the relative positions of the input force F_i, the output force F_o and the pivot point.
   a. First-class lever
   b. Second-class lever
   c. Third-class lever

8. Name three linear force transformers.

9. Define actual mechanical advantage in one or two sentences.

10. Distinguish between ideal mechanical advantage and actual mechanical advantage. When are they equal? When are they different? Which one is usually greater?

Student Challenge ─────────────────────────────────────

11. The efficiency of a force transformer is given by the equation,

$$\text{Eff} = \frac{\text{Work Out}}{\text{Work In}} \times 100\%,$$

or by the equivalent equation

$$\text{Eff} = \left(\frac{F_o \times D_o}{F_i \times D_i} \right) \times 100\%.$$

Knowing that $\text{IMA} = \dfrac{D_i}{D_o}$ and $\text{AMA} = \dfrac{F_o}{F_i}$, show that the efficiency is also given by

$$\text{Eff} = \frac{\text{AMA}}{\text{IMA}} \times 100\%.$$

12. A five-foot bar is used to raise a load. The pivot is four feet from the end of the bar where the input force is applied. In terms of the input and output forces, how much must the input end move down if the load is raised two inches?

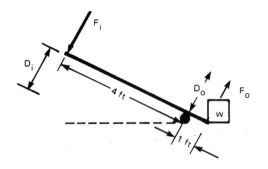

Hint: Use the equation that balances torques and the equation that equates "work in" with "work out." That is, use:

$$F_i \times L_i = F_o \times L_o$$

and

$$F_i \times D_i = F_o \times D_o$$

Solve for D_i, knowing that $D_o = 2$ inches.

MATH ACTIVITIES

Activity 1: Understanding Ratio and Proportion

Activity 2: Using Ratio and Proportion in Linear Mechanical Force Transformer Problems

MATH SKILLS LAB OBJECTIVES

When you complete these activities, you should be able to do the following:

1. **Define ratio. Give an example.**

2. **Define proportion. Give an example.**

3. **Show that the work equation, $F_i \times D_i = F_o \times D_o$, leads to a useful proportion, $\dfrac{F_o}{F_i} = \dfrac{D_i}{D_o}$.**

4. **Solve linear mechanical force transformer problems for mechanical advantage. Make use of ratios and proportions.**

LEARNING PATH

1. **Read the Math Skills Lab. Give particular attention to the Math Skills Lab Objectives.**

2. **Work the problems.**

ACTIVITY 1

Understanding Ratio and Proportion

Understanding how force transformers work is a lot easier if you understand ratio and proportion. A **ratio** is the quotient of one number divided by another, like "one divided by two" or $\frac{1}{2}$. In terms of letters, a ratio is one letter divided by another, like F_i/F_o. Think of a ratio as a fraction.

A **proportion,** on the other hand, is simply an equation that relates two equal ratios, such as $\dfrac{1}{2} = \dfrac{4}{8}$ or $\dfrac{F_i}{F_o} = \dfrac{D_o}{D_i}$.

WHAT ARE RATIOS?

Some ratios you've already studied in *Principles of Technology* are (1) angle measure in radians and (2) specific gravity. Notice that each ratio is **dimensionless.** That means that units always cancel in the final answer.

(1) *Angle measure in radians:*

$$\theta = \frac{D \text{ (arc length along circumference)}}{r \text{ (length of radius)}}$$

For example, if D = 1.5 cm and r = 1 cm,

$$\theta = \frac{D}{r} = \frac{1.5 \text{ cm}}{1.0 \text{ cm}} = 1.5 \text{ (Dimensionless!)}$$

The ratio is $^{1.5}\!/_1$ or 1.5. The angle (θ) is 1.5 radians or $1.5 \times 57.3° = 85.9°$.

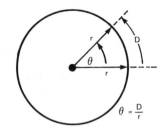

Fig. 1 Angle measure in radians.

(2) *Specific gravity (SG):*

$$SG = \frac{\text{Density of a Given Substance}}{\text{Density of Water}}$$

For example, the density of copper is 8.93 gm/cm³ and the density of water is 1.0 gm/cm³. The specific gravity (SG) of copper is a ratio given by:

$$SG = \frac{8.93 \text{ gm/cm}^3}{1.0 \text{ gm/cm}^3} = 8.93 \text{ (Dimensionless!)}$$

The ratio is $^{8.93}\!/_{1.0}$ (or 8.93, of course).

WHAT ARE PROPORTIONS?

A proportion connects two equal ratios. For example, each of the equations below involves a proportion. Notice that the left side and right side are each written as ratios.

$$\frac{1}{2} = \frac{4}{8}$$

This proportion is obvious. The fraction ½ is a ratio; the fraction ⅘ is a ratio. And they are equal. Each ratio means the same amount. The amount is simply stated in a different way. Often, a proportion (½ = ⅘) is read as "one is to two as four is to eight."

Now consider a different proportion problem.

$$\frac{1}{5} = \frac{a}{10}$$

This proportion has three numbers and one unknown. In this case, the unknown is the letter "a." You know that "a" can only be equal to 2 if the proportion is to be true. That's because 1 is to 5 as 2 is to 10 (⅕ = ²⁄₁₀).

Now consider a problem where there are two unknowns.

$$\frac{1}{2} = \frac{a}{b}$$

This proportion simply says that whatever a and b are, their ratio (ᵃ⁄b) must always be equal to ½. So if a = 5, then b = 10. Or if a = 10, then b = 20, and so on.

Now consider a proportion problem in which there are four unknowns.

$$\frac{a}{b} = \frac{c}{d}$$

This proportion has four letters. Their values are unknown. As it stands, the proportion simply tells us that "a is to b as c is to d." This isn't very useful. But if a, b, and c stand for something, the proportion tells you something important about how they are related and helps you solve for the value of the fourth letter (d).

HOW DO YOU USE RATIO AND PROPORTION IN FORCE TRANSFORMER EQUATIONS?

For all ideal force transformers, where no friction or resistance is present, it's true that: Work In = Work Out. Or, expressed another way,

$$F_i \times D_i = F_o \times D_o$$

where: F_i = input force
F_o = output force
D_i = distance input force moves
D_o = distance output force moves

The equation $F_i \times D_i = F_o \times D_o$ can be changed to one that involves a proportion and useful ratios, as follows:

$F_i \times D_i = F_o \times D_o$ — Always true if Work Input = Work Output.

$\dfrac{F_i \times D_i}{F_i \times D_o} = \dfrac{F_o \times D_o}{F_i \times D_o}$ — Divide each side by the product of $F_i \times D_o$. Cancel appropriately.

$\dfrac{D_i}{D_o} = \dfrac{F_o}{F_i}$ — A proportion results that involves ratios of forces and of distances.

The equation, $\dfrac{F_o}{F_i} = \dfrac{D_i}{D_o}$, is the same as $F_i \times D_i = F_o \times D_o$. It's just written in another form. But it's written as a proportion that says: "Output force F_o is to input force F_i as input displacement D_i is to output displacement D_o."

The ideal mechanical advantage (IMA) is always equal to the ratio D_i/D_o. But the proportion tells us that this ratio is the same as the ratio F_o/F_i.

Therefore, IMA $= \dfrac{D_i}{D_o}$ and IMA $= \dfrac{F_o}{F_i}$, too!

Remember that this is true only when Work In = Work Out—the case of *no friction*. Now solve a problem using ratio and proportion as they might show up in a force transformer problem.

Example A: *Ratio and Proportion*

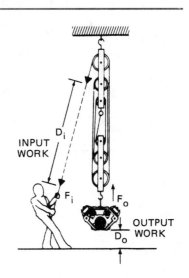

Given: The block and tackle shown is a force transformer. The important values are:

F_i = 100 lb
D_i = 6 ft
D_o = 1 ft

There is no friction.

Find: a. Output force F_o.
b. Force ratio F_o/F_i.
c. Displacement ratio D_i/D_o.
d. Ideal mechanical advantage.
e. Proportion between ratios F_o/F_i and D_i/D_o.

Solution: a. Since there's no friction, Work Out = Work In.

$$F_o \times D_o = F_i \times D_i$$

Rearrange the equation to solve for F_o.

$$F_o = \frac{F_i \times D_i}{D_o}$$

By substituting given values for F_i, D_i and D_o, we get:

$$F_o = \frac{100 \text{ lb} \times 6 \text{ ft}}{1 \text{ ft}} = \left(\frac{100 \times 6}{1}\right)\left(\frac{\text{lb·ft}}{\text{ft}}\right)$$

$$F_o = 600 \text{ lb}$$

b. Force ratio F_o/F_i:

$$\frac{F_o}{F_i} = \frac{600 \text{ lb}}{100 \text{ lb}} = \frac{6}{1}$$ (Notice that the ratio is dimensionless!)

The block and tackle is a "6-to-1" force amplifier.

c. Displacement ratio D_i/D_o:

$$\frac{D_i}{D_o} = \frac{6 \text{ ft}}{1 \text{ ft}} = \frac{6}{1}$$ (Notice, no units—no dimension.)

The trade-off is clear. To amplify force by a factor of 6, six times as much rope has to be moved at the input end.

d. Because there is no friction, the ideal mechanical advantage is either

$$\text{IMA} = \frac{F_o}{F_i} \text{ or IMA} = \frac{D_i}{D_o}, \text{ found in b and c.}$$

e. And now, let's figure the proportion. Since $F_i \times D_i = F_o \times D_o$, the proportion between the force ratio and the distance ratio becomes:

$$\frac{F_o}{F_i} = \frac{D_i}{D_o}, \text{ or } \frac{6}{1} = \frac{6}{1}$$

For this problem, each ratio is equal to $^6/_1$. This tells you that the "output force is to the input force" as the "input displacement is to the output displacement." Each one equals $^6/_1$, or "six times."

Now let's move to Activity 2. You can solve transformer problems where both ratio and proportion are used.

ACTIVITY 2

Using Ratio and Proportion in Linear Mechanical Force Transformer Problems

For this activity, you'll solve problems that commonly occur in industrial, commercial and construction applications of linear force transformers.

Use the equations for mechanical advantage and the relationships of work input to work output to solve the problems. Assume that the transformers are **IDEAL,** unless stated otherwise. The equations and relationships are as follows:

1. Work Input = Work Output
$$F_i \times D_i = F_o \times D_o \quad \text{(no friction)}$$

2. Two important ratios:
$$\text{IMA} = \frac{D_i}{D_o} \text{ and IMA} = \frac{F_o}{F_i}$$

3. An important proportion:
$$\frac{F_o}{F_i} = \frac{D_i}{D_o} \qquad \text{(no friction)}$$

Problem 1: Given: A "front loader" is a type of earth-moving machine that has a scoop bucket, known as a "loader bucket," attached to the front of the machine. The bucket can be rotated forward and backward by a hydraulic cylinder. The bucket pivots on a lifting-arm mechanism. This bucket pivot forms a first-class lever. Assume no friction.

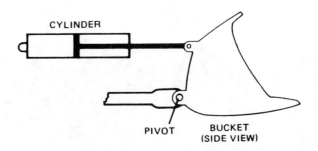

CYLINDER

PIVOT   BUCKET
(SIDE VIEW)

Find: a. Draw a picture of the forces, the pivot and the displacements as they apply to the front loader.

b. The bucket lever arm has an ideal mechanical advantage of 4. The hydraulic cylinder applies 20,000 lb of force to the input end of the lever. What's the output force applied by the lever on the load?

Note: $IMA = \dfrac{D_i}{D_o} = 4$ and Work In = Work Out.

Solution:

Problem 2: Given: A portable engine hoist uses an air cylinder to raise and lower the hoist arm (a third-class lever). The height the load is raised (D_o) is more important than the lifting capacity (F_o). The hydraulic cylinder exerts a 6000-lb force (F_i) on the arm. The hoist arm has a 1000-lb lifting force (F_o) at the lifting hook. Assume no friction, so that $IMA = AMA = F_o/F_i = D_i/D_o$.

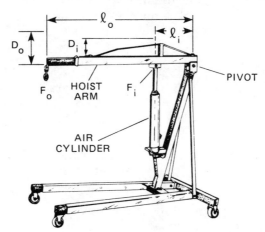

ℓ_o ℓ_i

D_o D_i PIVOT

F_o HOIST F_i
         ARM

AIR
CYLINDER

Find: a. The mechanical advantage of the hoist (either IMA or AMA).

b. The distance (D_o) the lifting hook raises the load when the piston arm in the cylinder extends outward 6 inches from the closed position.

Solution:

Problem 3: Given: A dump truck has a lift cylinder that exerts 7 tons of force at the front of the 12-foot-long dump bed. The pivot point (hinge) is at the foot of the dump bed.

Find:

a. Draw a simple picture that locates the load, applied force and pivot. (**Hint:** It's a second-class lever.)

b. Assuming the load is concentrated at the center of the dump bed, find the weight this truck can dump when fully loaded.

Solution:

Problem 4: Given: A garage door spring exerts a force of 150 lb while doing 300 ft·lb of work on the frictionless lever mechanism of the door. The ideal mechanical advantage of this mechanism is 0.33.

Find: The output force required to raise the door 6 ft (under ideal conditions). Remember that work input equals work output if friction/resistance is ignored.

Note: In practice, garage doors use springs and lever mechanisms to create a *counterbalance*. (A counterbalance is a condition that exists when one weight balances another weight.) Under these conditions, the torques are equal and opposite in direction. The door is in equilibrium. A slight input force changes this condition. The door opens quite easily.

Solution:

Problem 5: Given: An auto bumper jack is 80% efficient. To raise the wheel of a certain auto, an 80-pound input force moves down 12 inches. The jack applies a force that raises the wheel one inch.

Find:

a. The load force, if Work Input = Work Output (ideal conditions).

b. What is the AMA if efficiency is 80%?

c. What is the actual force applied to the load?

Note: Eff $= \dfrac{\text{AMA}}{\text{IMA}} \times 100\%$ and

IMA $= \dfrac{D_i}{D_o}$ (ideal conditions).

Solution:

Problem 6: Given: The diagrams below of two pulley systems.

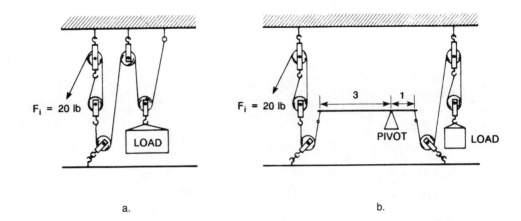

a.   b.

Find: a. The load that can be lifted in each arrangement if the input force is 20 lb. Assume no friction.

b. The distance the load will be lifted in each case if the input force is applied over a 24-in. distance. Justify your answer in one or two sentences.

Solution:

Linear Mechanical Transformers: The 'Come-along' Winch

LAB OBJECTIVES

When you've finished this lab, you should be able to do the following:

1. *Describe how a come-along winch works. Tell how it's made up of three simple machines. Explain which kind of transformer each machine is.*

2. *Measure input force, input displacement, output force and output displacement of a come-along winch in operation.*

3. *Use a come-along winch. Determine its ideal mechanical advantage, actual mechanical advantage and efficiency.*

LEARNING PATH

1. *Preview the lab. This will give you an idea of what's ahead.*

2. *Read the lab. Give particular attention to the Lab Objectives.*

3. *Do lab, "Linear Mechanical Transformers: The 'Come-along' Winch."*

MAIN IDEAS

- *A small mechanical input force applied through a linear distance results in the output of a much larger force delivered over a much shorter distance.*

- *The "come-along" winch is a combination of three simple machines.*

- *Resistance reduces the efficiency of any machine.*

Technology isn't always complicated. It often takes a simple machine and combines it with other simple machines to produce a device that does useful work.

The winching device called the "come-along" is just such a machine. This device is shown in Figure 1. It's useful.

For example, ranchers and farmers use come-alongs to tighten wire fences. Auto mechanics use them to lift engines out of cars. Four-wheel-drive enthusiasts often use this device to get their cars out of deep mud. This is because the come-along is an easy means to develop a large force from a much smaller force.

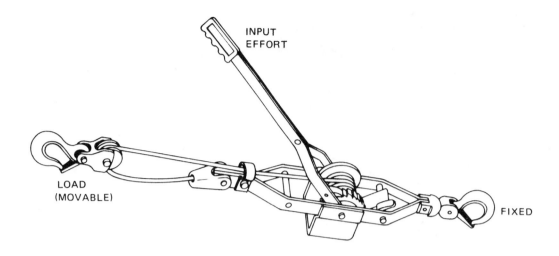

Fig. 1 The "come-along" winch.

The come-along combines three simple machines. Each of the machines has its own mechanical advantage

One of the simple machines is the *input lever.* This is a second-class lever. Its pivot (or fulcrum) is placed at the center of rotation of the second simple machine—a *wheel and axle.* The axle is a drum that takes up the steel cable. This drum has a smaller diameter than the gear-tooth wheel that drives it. The IMA is the ratio of the two diameters. The *pulley* on the come-along cable is the third simple machine. When you use the pulley on the output hook, the load is supported by a double-part cable. This "two-part line" is better than a single-part cable. The more lines supporting the load, the greater the mechanical advantage. It's the number of lines *to the load* that makes the difference. In Figure 1 you can see that the IMA of the pulley is 2.

The total mechanical advantage is the product of the mechanical advantage of each simple machine. In this lab, you'll use a come-along winch to lift a known load.

LABORATORY ───────────────────────────

EQUIPMENT

    Steel rule
    Heavy-duty support stand
    Large weight set, total 10 kg
    Spring scale, 5-lb (20-N) or 25-lb (100-N) capacity
    Come-along winch
    Meterstick/yardstick
    Protractor

PROCEDURES

Part 1: *Getting to Know the Come-along and Making Basic Measurements*

1. Look carefully at the come-along winch. Figure 2 shows the important details for the winch. Make note of the location of various levers, pulleys, wheels and gear teeth. Refer to this figure for measurements described in Steps 2 through 6.

2. Measure the distance (ℓ_i) from the handle to the fulcrum. See Figure 2. Record this measurement in Data Table 1.

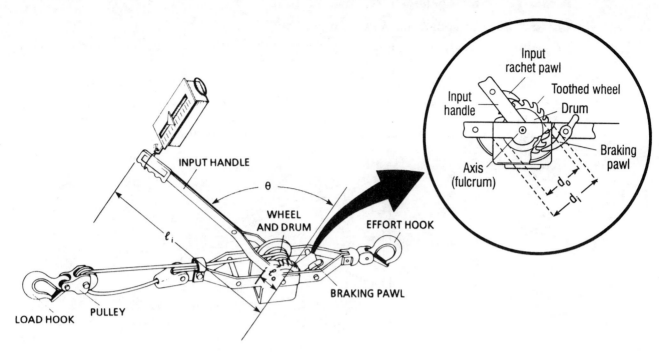

Fig. 2 Details of come-along winch.

3. Measure the distance (ℓ_o) from the ratchet pawl to the fulcrum. See Figure 2. Record in Data Table 1.

4. Measure the diameter (d_i) of the toothed wheel. See inset in Figure 2. Record in Data Table 1.

5. Measure the diameter (d_o) of the drum on which the cable is wrapped. See inset. Record in Data Table 1.

6. Count the number of teeth on the toothed wheel, N. Then count the number of clicks heard with each full stroke, n. The angle of displacement θ the wheel is turned with each full stroke is then $\theta = {}^n\!/_N \times 2\,\pi$ (in radians). Record θ in Data Table 1.

DATA TABLE 1: BASIC MEASUREMENTS

Refer to Figure 2 for meaning of symbols.
Lever   Distance from handle to fulcrum . . . . ℓ_i = _____   Distance from input ratchet pawl to fulcrum . . . . ℓ_o = _____   **Wheel and axle**   Diameter of toothed wheel . . . . d_i = _____   Diameter of drum . . . . d_o = _____   **Angle of one complete stroke** . . . . θ = _____

7. Disengage the braking pawl. Pull out the full amount of steel cable wrapped around the drum.

8. Measure the total length of pull that's possible with this device. Record that length on your data sheet as the "pull length."

9. Have a partner hold the pulling hook so the cable is straight. Then find how many revolutions of the wheel/drum it takes to reel in all the cable. Record on your data sheet the "revolutions needed to reel in hook."

1. Set up your apparatus, as shown in Figure 3.

2. Unreel enough cable to allow the weight hanger to rest on the base of the support stand.

3. Add a total of 10 kg of slotted weights to the hanger. Since the hanger weighs 1 kg, the total weight attached to the hook is 11 kg.

4. Apply tension to the cable by turning the cable drum. You have applied the correct amount of tension when the weight hanger is almost completely supported by the cable.

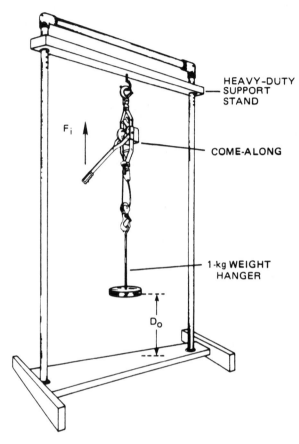

Fig. 3 Lab setup for come-along winch.

Note: The next few steps require at least two or three people working together.

5. Attach the spring scale to the input handle of the winch.

6. While one person holds the winch steady, another person should apply a constant force to the input lever by pulling on the spring scale. The line of action of force applied to the input lever must be at 90° (perpendicular) to the lever. This is shown in Figure 4.

   Read the input force (F_i) measured by the spring scale at the **beginning** of the pull while the handle is just moving. Then ignore the spring scale. Continue pulling (with your hand) through the rest of the stroke.

   Record the input force (F_i) measured by the spring scale in Data Table 2, opposite "Stroke No. 1."

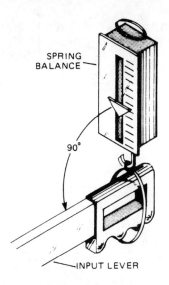

SPRING
BALANCE

90°

INPUT LEVER

Fig. 4 Measuring input force.

7. After the application of one full stroke, measure the displacement distance (D_o) that the load has been raised. Record that distance on your data sheet as D_o.

8. Repeat Step 6 about 10 more times, measuring F_i each time. Record F_i in Data Table 2 for each stroke. You need not measure the distance (D_o) the weight moves up each time.

DATA TABLE 2: LIFTING A LOAD

Stroke No.	Applied Force (F_i)	Stroke No.	Applied Force (F_i)
1.		6.	
2.		7.	
3.		8.	
4.		9.	
5.		10.	

Calculations

1. Find the ideal mechanical advantage of the come-along winch. To do this, find the ideal mechanical advantage for each part. Then multiply your answers as outlined below. Refer to Data Table 1 for data.

$$\text{IMA (lever)} = \frac{\ell_i}{\ell_o} = \frac{(\quad)}{(\quad)} = \underline{\quad} .$$

$$\text{IMA (wheel \& drum)} = \frac{d_i}{d_o} = \frac{(\quad)}{(\quad)} = \underline{\quad} .$$

IMA (pulley) = 2 (given)

IMA (winch) = IMA (lever) × IMA (wheel & drum) × IMA (pulley)
IMA (winch) = (_____) × (_____) × (2)
IMA (winch) = _____ (from product of individual IMAs)

2. Find the total arc length the applied force F_i moves through in one stroke. This should be the input displacement D_i. Refer to Figure 2 and Data Table 1 for data.

   D_i = Lever arm (ℓ_j) × angle moved through (θ) in radians.

   D_i = (_____) × (_____)

   D_i = _____

3. Find the ideal mechanical advantage of the winch from the equation, $IMA = D_i/D_o$. Use the D_i just calculated and D_o from Step 7 under Part 2.

   $$IMA = \frac{D_i}{D_o} = \frac{(\quad)}{(\quad)} = \underline{\quad} \quad \text{(from overall performance)}$$

4. Find the actual mechanical advantage (AMA) of the winch.

   a. First, find an "average" applied force F_i. To do this, add all of the "applied forces" in Data Table 2. Divide by the number of entries.

   $$Avg\ F_i = \frac{Sum\ of\ all\ F_i}{Number\ of\ F_i} = \frac{(\quad)}{(\quad)} = \underline{\quad}$$

   b. Next, find the output force F_o developed by the winch. Convert the load (11 kg) to a weight in newtons or pounds, depending on the units read on the spring scale for F_i.

   $$F_o(N) = 11\ \cancel{kg} \times \frac{9.8\ N}{1\ \cancel{kg}} = 107.8\ N \qquad \text{(Cancel units of kg.)}$$

   or

   $$F_o(lb) = 11\ \cancel{kg} \times \frac{2.2\ lb}{1\ \cancel{kg}} = 24.2\ lb \qquad \text{(Cancel units of kg.)}$$

   c. Now find the actual mechanical advantage.

   $$AMA = \frac{F_o}{Avg\ F_i} = \frac{(\quad)}{(\quad)} = \underline{\quad}$$

5. Find the efficiency of the come-along winch, using the results of Steps 3 and 4.

   $$Eff = \frac{AMA}{IMA} \times 100\% = \frac{(\quad)}{(\quad)} \times 100 = \underline{\quad} \%$$

WRAP-UP

1. How do the two ideal mechanical advantages found in Steps 1 and 3 of the Calculations compare? Should they be the same?

2. Why is the efficiency in Step 5 of Calculations less than 100%?

Student Challenge

1. For the come-along winch, can you explain why the overall mechanical advantage might be greater while the cable is reeled out than when it's almost reeled in? Which of the "transformers"—lever, wheel and drum, or pulley—is involved?

2. Identify the procedure for reasonable care and maintenance you'd follow to ensure high efficiency and a long life for the come-along winch.

Linear Force Transformers: The Screw

LAB OBJECTIVES

When you've finished this lab, you should be able to do the following:

1. *Identify the main parts of the simple machine known as the screw.*
2. *Find the ideal mechanical advantage of a given pipe vise.*
3. *Measure input and output forces on a pipe clamp. Find the actual mechanical advantage.*

LEARNING PATH

1. *Preview the lab. This will give you an idea of what's ahead.*
2. *Read the lab. Give particular attention to the Lab Objectives.*
3. *Do lab, "Linear Force Transformers: the Screw."*

MAIN IDEAS

- *The inclined plane is a common simple machine that takes many forms, one of which is the screw.*
- *Ideal mechanical advantage of a pipe clamp or screw is given by the ratio of "distance moved by input force" to "pitch of the screw."*
- *Actual mechanical advantage of a screw (ratio of force out to force in) is much less than the ideal mechanical advantage, for most screws.*
- *Friction greatly reduces the efficiency of a screw device. Lubricating screw threads increases efficiency.*

Inclined planes are simple machines that have been used for more than 4000 years. Today, inclined planes are used to move heavy loads to a higher level just as they were thousands of years ago.

In addition, the tapered shape of the inclined plane is used in mechanical cutting tools. These tools include scissors, chisels, wedges and knives. The inclined plane is used in other ways and can be applied in different forms.

For example, a cam is a continuous inclined plane. A wedge is two inclined planes, back to back. A screw is an inclined plane that's been wrapped around a cylinder or cone.

Screws are used as fasteners and as working components in machines. A home trash compactor uses a set of screws to crush trash. Some overhead garage door openers use a long screw to raise and lower the door.

A grain grower commonly uses a device called a "grain auger" (a screw-shaped device) to move grain from a truck into a grain silo. Some types of photocopiers use a screw mechanism as part of the paper-feed system. The height of an office chair is often adjusted by a screw mechanism. Some types of automotive jacks use a screw as part of the lifting mechanism.

Figure 1a identifies the main parts of a screw. Figures 1b, 1c and 1d show the use of a screw in several of the devices mentioned.

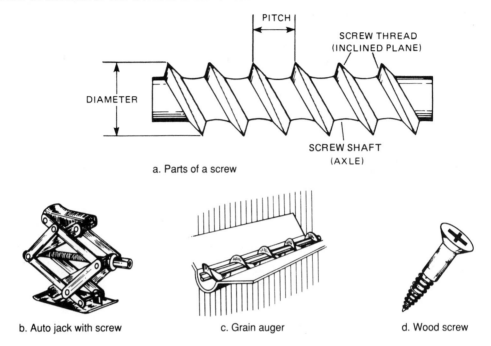

a. Parts of a screw

b. Auto jack with screw   c. Grain auger   d. Wood screw

Fig. 1 The screw.

HOW DO YOU FIND THE MECHANICAL ADVANTAGE OF THE SCREW?

To find a screw's ideal mechanical advantage, compare the ratios of the input displacement to the output displacement. The input displacement D_i is the circumference of the turning circle $(2\pi r)$ through which the input moves.

The radius (r) of that circle is the distance from the center of the axis of rotation to the point of the input force (F_i). The output displacement D_o is equal to the pitch of the screw. That's the distance moved along the shank of the screw after one full turn.

Figure 2 shows an automotive-type, tripod screw jack. All of the parameters needed to find ideal mechanical advantage, actual mechanical advantage and efficiency are labeled.

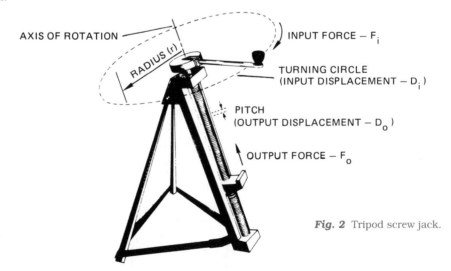

Fig. 2 Tripod screw jack.

Equation 1 shows the formula for finding the ideal mechanical advantage (IMA) of a screw.

$$\text{IMA} = \frac{\text{Circumference of Turning Circle}}{\text{Pitch of Screw Thread}} = \frac{2\pi r}{P} = \frac{D_i}{D_o}$$ **Equation 1**

Equation 2 shows how to find the actual mechanical advantage (AMA) of a screw.

$$\text{AMA} = \frac{\text{Output Force}}{\text{Input Force}} = \frac{F_o}{F_i}$$ **Equation 2**

The percent efficiency of the screw is given by Equation 3.

$$\text{Eff} = \frac{\text{AMA}}{\text{IMA}} \times 100\%$$ **Equation 3**

LABORATORY

EQUIPMENT
Screw and disk assembly
Steel rule
Single-sheave pulley
Meterstick/yardstick
Spring scale, 8-oz (2.5-N) or 16-oz (5-N) capacity
Small weight set, total 1 kg
Small weight hanger, 50-g type
Cord, minimum 30-lb test

PROCEDURES

Part 1: Making Initial Measurements on the Pipe Clamp

1. Measure and record the dimensions asked for in Data Table 1. Refer to Figures 1a and 3.

DATA TABLE 1

Diameter of Input Disk	d_i = _____
Diameter of Screw Shaft	d_s = _____
Number of Turns Necessary to Travel the Total Span of Screw	n = _____
Length of the Total Span of Screw	L = _____
Pitch of Screw	n/L = _____

2. What are the two simple machines used in the pipe clamp? Write your response on your data sheet.

Part 2: Measuring Input/Output Forces

1. Assemble and set up the lab apparatus, as shown in Figure 3. Wind a cord around the disk. Attach the spring scale. Connect the cord pulley and weight hanger, as shown. Also, place a single mark on the disk, as shown.

2. Be sure that you've wound the cord around the disk in the proper direction so that when you pull on the spring scale, the screw will turn and cause the weight hanger to *rise*.

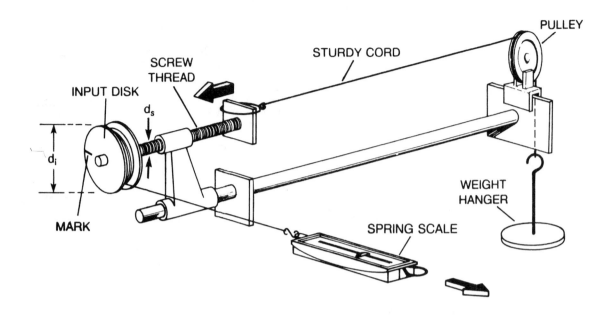

Fig. 3 Lab setup for screw and disk assembly.

3. If the weight hanger alone does not take the slack out of the cord, add sufficient weight. Pull on the spring scale along a straight line (parallel to the faces of the disk) with a constant force. When the mark on the disk points straight down, follow directions *a*, *b*, and *c* below.

   a. In Data Table 2, record the force indicated on the spring balance (Input Force F_i) and the total weight pulling down on the cord (Output Force F_o). Record these as "Reading 1."

   b. Approximately double the weight on the weight hanger and again record F_i and F_o values, as in Step a, in Data Table 2, row "Reading 2."

   c. Increase the weight until you've measured and recorded F_i and F_o for three more trials: Readings 3, 4 and 5.

DATA TABLE 2

Reading	Input Force (F_i)	Output Force (F_o)
1		
2		
3		
4		
5		

1. Find the ideal mechanical advantage (IMA) of the screw and disk assembly.

$$IMA = \frac{(\pi)(\text{Diameter of Disk})}{\text{Pitch of Screw}} = \frac{(3.14)(\quad)}{(\quad)}$$

2. Find the actual mechanical advantage (AMA) of the screw and disk assembly. The AMA for each set of readings recorded in Data Table 2 is found from the following equation:

$$AMA = \frac{F_o}{F_i}$$

Record the value of AMA for each set of readings in Data Table 3.

DATA TABLE 3

Reading	AMA
1	
2	
3	
4	
5	
Average	

Find the average AMA by adding the separate values of AMA. Then divide by 5, the number of values. Record in Data Table 3.

3. Find the efficiency of the screw and disk assembly using the average AMA (Step 2) and IMA (Step 1).

$$Eff = \frac{AMA}{IMA} \times 100\% = \underline{\quad}\%$$

WRAP-UP

1. Why did the screw assembly have less than 100% efficiency?
2. If the screw pitch were increased (that is, less threads per unit of length), would there be a change in the ideal mechanical advantage? If so, how would it change?
3. What can you do to increase the efficiency of a screw?

Student Challenge ───

1. How would you find input work for the screw assembly? What two quantities need to be found?
2. How would you find output work for the screw assembly? What two quantities need to be found?
3. Knowing input work and output work, can you find out if any input work has been lost? How? Where did the loss go? How can you reduce this loss?
4. Suppose you wished to raise the weight hanger that's connected to the screw at a speed of 1 ft/min. What would the rpm of the disk have to be?

Review

1. View and discuss the video, "Force Transformers in Linear Mechanical Systems."

2. Review the Objectives and Main Ideas of the print materials in this subunit.

3. Your teacher may give you a test over Subunit 1, "Force Transformers in Linear Mechanical Systems."

SUBUNIT 2

Force Transformers in Rotational Mechanical Systems

SUBUNIT OBJECTIVES

When you've finished reading this subunit and viewing the video, "Force Transformers in Rotational Mechanical Systems," you should be able to do the following:

1. Explain the relationship between the input work and output work for rotational force transformers.

2. Find the mechanical advantage and efficiency of a wheel and axle.

3. Find the mechanical advantage and efficiency of a belt-drive system.

4. Find the mechanical advantage and efficiency of a gear-drive system.

5. List or identify various rotational force transformers.

6. Measure the mechanical advantage of a rotational force transformer.

7. Identify workplace applications where technicians use rotational force transformers.

LEARNING PATH

1. Read this subunit, "Force Transformers in Rotational Mechanical Systems." Give particular attention to the Subunit Objectives.

2. View and discuss the video, "Force Transformers in Rotational Mechanical Systems."

3. Participate in class discussions.

4. Watch a demonstration about rotational mechanical force transformers.

5. Complete the Student Exercises.

MAIN IDEAS

*   The equations that apply to linear force transformers can be adapted to help you understand how rotational force transformers operate.

*   Rotational "force" transformers generally amplify (increase) force.

- **Work In equals Work Out in 100%-efficient rotational force transformers. Work Out is less than Work In when resistance is present.**

- **A "wheel and axle" amplifies force and has many practical uses.**

- **Ideal mechanical advantage for a wheel and axle is given by the equations, IMA = D_i/D_o = r_i/r_o. Actual mechanical advantage for a wheel and axle is given by the equation, AMA = F_o/F_i.**

- **Belt drives, gear drives and disk drives use similar methods to achieve trade-offs between torque and speed.**

- **Rotational force transformers are used throughout industry to provide a mechanical advantage.**

In this subunit, you'll study some useful *rotational* force transformers. To simplify the discussion, we'll say the force transformers are frictionless. This makes calculations easier. But remember, **real** force transformers are not 100% efficient.

WHAT'S THE RELATIONSHIP BETWEEN INPUT WORK, OUTPUT WORK AND TORQUE TRANSFORMERS?

From the "Overview" and "Subunit 1," you know that a force transformer is a coupling device. You know that work done on the coupling device by an input force changes to work done on a load by an output force.

The coupling device or force transformer can increase either output force or output displacement—but not both.

You also know that work in linear mechanical systems is described by the following relationship:

$$\text{Work In} = F_i \times D_i$$

In a rotational mechanical system, the input to a rotational force transformer is the energy produced by work done at the input or source. This is **rotational work.**

Rotational work equals the **input torque applied** (T_i) times the **input angle** (θ_i) through which the force moves.

$$\text{Work In} = T_i \times \theta_i$$

A rotational force transformer uses input energy to do rotational work on the load. Rotational output work equals the **output torque** (T_o) times the **output angle** (θ_o) through which the force moves.

$$\text{Work Out} = T_o \times \theta_o$$

If the transformer is 100% efficient, work output equals work input.

$$\text{Work In} = \text{Work Out}$$

To show this relationship for rotational mechanical systems, the equation is:

$$T_i \times \theta_i = T_o \times \theta_o \qquad \textbf{Equation 7}$$

where: $T_i = F_i \times r_i$ = input torque

θ_i = angle the input force (F_i) moves through (in radians)

$T_o = F_o \times r_o$ = output torque

θ_o = angle the output force (F_o) moves through (in radians)

r_i = lever arm for input force

r_o = lever arm for output force

The value of θ was discussed in Unit 2, *Work*. Figure 7-14 shows how to find θ. In Figure 7-14, the displacement (D) along the rim of the wheel is related to the radius (r) of the wheel and the angle through which the wheel rotates (θ).

The value of θ equals the **ratio** of the rim distance D to the radius r of the wheel. So $\theta = D/r$. The angle is measured in **radians.**

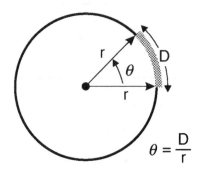

$$\theta = \frac{D}{r}$$

Fig. 7-14 Angular distance.

Let's apply this relationship to Equation 7. Substitute for T and θ in the work equation.

$$\text{Work In} = \text{Work Out}$$

$$T_i \times \theta_i = T_o \times \theta_o$$

where:

$$T_i = F_i \times r_i$$
$$T_o = F_o \times r_o$$
$$\theta_i = D_i/r_i$$
$$\theta_o = D_o/r_o$$

Therefore,

$$(F_i \times \cancel{r_i}) \times \left[\frac{D_i}{\cancel{r_i}}\right] = (F_o \times \cancel{r_o}) \times \left[\frac{D_o}{\cancel{r_o}}\right]$$

Cancel like terms to obtain:

$$F_i \times D_i = F_o \times D_o$$

So the relationship that "work input equals work output" in a rotational mechanical system leads to the work equation relating forces and displacements. This is true in linear mechanical systems. Therefore, you can form the same ratios for ideal and actual mechanical advantage that you did with linear systems:

$$IMA = \frac{D_i}{D_o}$$

$$AMA = \frac{F_o}{F_i}$$

As a result, the equations for ideal mechanical advantage and actual mechanical advantage remain the same. They can apply to either linear or rotational force transformers.

WHAT'S A WHEEL AND AXLE?

Suppose you have two wheels of different radii, r_i and r_o, fastened together about a shaft. (Such a system is shown in Figure 7-15.) An input force F_i applied at a point on the rim of the larger wheel of radius r_i causes that point to move through an arc of distance D_i.

At the same time, the smaller wheel of radius r_o moves through an arc of distance D_o, while putting a force F_o on the load. Note that D_o and D_i are *not* equal because the radii of the two wheels are different.

But the wheels are mechanically connected so that the angle through which they turn is the same. The two wheels acting together amplify force F_i into force F_o.

You can combine the equations for ideal mechanical advantage (IMA = D_i/D_o) and the equations that connect rim distance with angle ($\theta_i = D_i/r_i$ and $\theta_o = D_o/r_o$). When they are combined, the ideal mechanical advantage (IMA) is equal to a simple ratio—the ratio of the large wheel radius r_i to the small wheel radius r_o.

Thus, Equation 8 sums up the ideal mechanical advantage for a rotational force transformer.

$$IMA = \frac{D_i}{D_o} = \frac{r_i}{r_o} \qquad \textbf{\textit{Equation 8}}$$

Remember that the actual mechanical advantage (AMA = F_o/F_i) will always be less than the ideal mechanical advantage (IMA) when resistance due to friction is present.

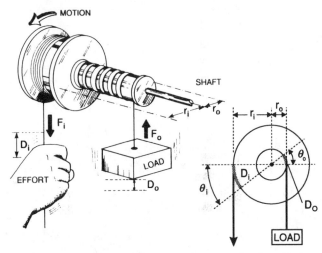

Fig. 7-15 A rotational force transformer—wheel and axle.

The machine or force transformer shown in Figure 7-15 exists in numerous forms. It's usually called a wheel and axle. The most common form of the wheel and axle is the mechanical chain hoist used in garages to take engines out of automobiles.

Figure 7-16 shows a picture and a schematic of a chain hoist. The chain hoist is made of two wheels (pulleys) rigidly connected and mounted on the same shaft. Both wheels rotate together.

The larger of the two wheels has an endless chain wrapped around its rim. The rim has notches. Notches keep the chain from slipping when an input force is applied to the wheel by pulling on the chain. The smaller wheel has the same type notches to keep the load chain from slipping as it rotates over the small wheel rim. A mechanical advantage is gained by using a small input force that results in a large output force.

Many other devices work like the wheel and axle or chain hoist. The wind-up winch on a boat trailer—and fishing reels with wind-up drums—are all wheel-and-axle devices. The wind-up handle replaces the larger wheel. The drum replaces the smaller wheel. Industry has adapted the wheel and axle to get a mechanical advantage for many machine uses. Among these are the cable winch, the overhead crane and many other lifting machines.

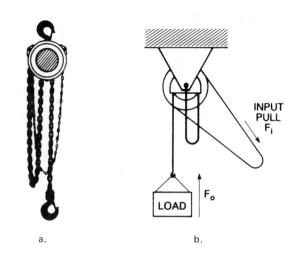

Fig. 7-16 Chain hoist.

Table 7-1 sums up useful formulas that apply to how wheel-and-axle force transformers work.

TABLE 7-1. FORMULAS FOR A WHEEL AND AXLE

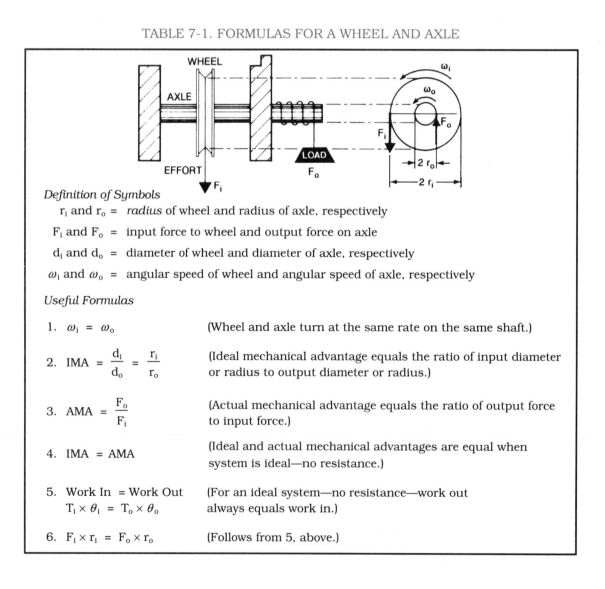

Definition of Symbols

r_i and r_o = *radius* of wheel and radius of axle, respectively

F_i and F_o = input force to wheel and output force on axle

d_i and d_o = diameter of wheel and diameter of axle, respectively

ω_i and ω_o = angular speed of wheel and angular speed of axle, respectively

Useful Formulas

1. $\omega_i = \omega_o$ (Wheel and axle turn at the same rate on the same shaft.)

2. $IMA = \dfrac{d_i}{d_o} = \dfrac{r_i}{r_o}$ (Ideal mechanical advantage equals the ratio of input diameter or radius to output diameter or radius.)

3. $AMA = \dfrac{F_o}{F_i}$ (Actual mechanical advantage equals the ratio of output force to input force.)

4. $IMA = AMA$ (Ideal and actual mechanical advantages are equal when system is ideal—no resistance.)

5. Work In = Work Out
   $T_i \times \theta_i = T_o \times \theta_o$ (For an ideal system—no resistance—work out always equals work in.)

6. $F_i \times r_i = F_o \times r_o$ (Follows from 5, above.)

Work through Example 7-G. This example will help you understand how a wheel and axle works as a force transformer. Use formulas in Table 7-1 as needed.

Example 7-G: *A Chain Hoist and Its Mechanical Advantage*

Given: A force of 50 lb is applied on the input chain of a chain hoist. The chain hoist has a 12-in.-diameter input wheel and a 4-in.-diameter load wheel.

Find: a. The load lifted by the 50-lb input force.

        b. The mechanical advantage of the chain hoist.

Solution: a. Find the load lifted by the chain hoist. That is, find F_o. Assume a ***frictionless*** chain hoist. So you know that:

Work In = Work Out

$T_i \times \theta_i = T_o \times \theta_o$

where: T_i = input torque

        θ_i = input angle moved through by input force F_i

        T_o = output torque

        θ_o = output angle moved through by output force F_o

Since the two wheels are mounted on the same shaft, both turn the same amount at the same time. Therefore, $\theta_i = \theta_o$. This means that $T_i = T_o$, or

Torque In = Torque Out

$F_i \times r_i = F_o \times r_o$

where: r_i = input wheel radius

        r_o = output wheel radius

To solve for F_o, rearrange the equation to give:

$$F_o = \frac{F_i \times r_i}{r_o}$$

Substitute values, where F_i = 50 lb, $r_i = \tfrac{1}{2}(d_i)$ = 6 in., and $r_o = \tfrac{1}{2}(d_o)$ = 2 in.

$$F_o = \frac{50 \text{ lb} \times 6 \text{ in.}}{2 \text{ in.}}$$

$$F_o = \left(\frac{50 \times 6}{2}\right)\left(\frac{\text{lb·in.}}{\text{in.}}\right)$$

$$F_o = 150 \text{ lb}$$

    b. Find the mechanical advantage of the chain hoist. From Equation 8, the ideal mechanical advantage involves the ratio of the two radii. The actual mechanical advantage involves a ratio of output and input forces. Use Formulas 2 and 3 in Table 7-1.

$$\text{IMA} = \frac{r_i}{r_o} = \frac{6 \text{ in.}}{2 \text{ in.}} = 3$$

$$\text{AMA} = \frac{F_o}{F_i} = \frac{150 \text{ lb}}{50 \text{ lb}} = 3$$

Note: IMA = AMA because you assumed that the chain hoist was 100% efficient.

HOW DO YOU FIND THE EFFICIENCY OF A ROTATIONAL FORCE TRANSFORMER?

The efficiency of a rotational force transformer is given by the same formula as that for a linear force transformer. Thus, $\text{Eff} = \dfrac{\text{IMA}}{\text{AMA}} \times 100\%$. Use this equation (in Example 7-H) to find the efficiency of the chain hoist described in Example 7-G.

Example 7-H: *Efficiency of a Chain Hoist*

Given: The chain hoist problem of Example 7-G. The values given for the chain hoist were as follows:

Input Force F_i = 50 lb
Input Wheel Radius r_i = 6 in.
Output Wheel Radius r_o = 2 in.

From Formula 2 in Table 7-1:

$$\text{Ideal Mechanical Advantage (IMA)} = \frac{r_i}{r_o} = \frac{6 \text{ in.}}{2 \text{ in.}} = 3$$

Find:
a. The actual mechanical advantage of the hoist, if output force F_o = 140 lb.
b. The efficiency of the hoist under these conditions.

Solution:

a. $\text{AMA} = \dfrac{F_o}{F_i} = \dfrac{140 \text{ lb}}{50 \text{ lb}} = 2.8$

b. $\text{Eff} = \dfrac{\text{AMA}}{\text{IMA}} \times 100\% = \dfrac{2.8}{3} \times 100\% = 93\%$

> **Note:** The ideal mechanical advantage (IMA) involves the fixed values of the radii. The actual mechanical advantage (AMA) involves forces that depend on the presence or absence of friction or internal resistance. That's why the output force F_o = 150 lb found in the ***frictionless*** chain hoist in 7-G was reduced to F_o = 140 lb in the ***actual*** chain hoist, in 7-H. In Example 7-H, there was enough friction to reduce the 150-lb output force to 140 lb.

WHAT ARE DRIVE SYSTEMS?

Consider rotational mechanical systems. Force transformers called "drive systems" are used when you want **changes in torque and angular speed.**

Three of the most common of these **torque** transformers are (1) belt-drive systems, (2) disk-drive systems and (3) gear-drive systems. Let's examine the belt-drive system first. We'll use what you already know about mechanical advantage from the wheel and axle.

HOW DO BELT DRIVES WORK?

The wheel and axle uses two wheels of different radii fastened together on the same shaft or axle. An input force at the rim of the larger wheel creates an output force at the rim of the smaller wheel. The output force—larger than the input force—provides a mechanical advantage when moving a heavy load.

In a belt drive, the wheels are on **different axles.** (See Figure 7-17.) The wheels—or **pulleys,** as they are often called—are rigidly

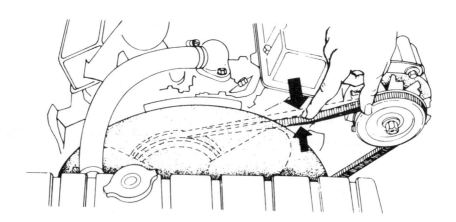

Fig. 7-17 A belt-drive system.

fastened to the two axles or shafts. When the input shaft rotates, the input pulley moves with it. This moves the belt. Belt movement causes the output pulley to turn.

Look at the drawing of a belt-drive system in Table 7-2. If there's no belt slippage, then:

a.   The distance D_i moved by the belt along the input wheel always **equals** the distance D_o moved along the output wheel.

b.   The **smaller** wheel always turns **faster** than the **larger** wheel.

c.   If the **input torque** is applied to the **smaller** wheel, the **output torque** on the **larger** wheel is **increased.**

d.   If the **input torque** is applied to the **larger** wheel, the **output torque** on the **smaller** wheel is **decreased.**

WHAT ARE SOME USEFUL FORMULAS FOR BELT-DRIVE SYSTEMS?

Table 7-2 sums up the formulas that apply to belt-drive systems. With these formulas, you can find angular speed of the shafts and torques on the pulleys. You can also find the ideal or actual mechanical advantages of the belt drive.

TABLE 7-2. FORMULAS FOR A BELT-DRIVE SYSTEM

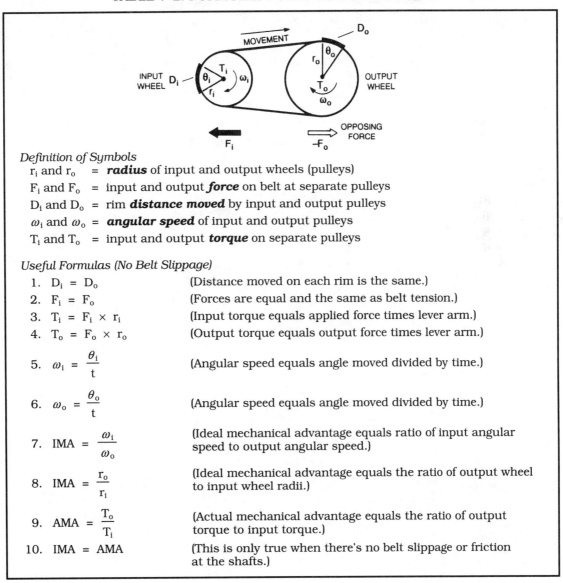

Definition of Symbols

r_i and r_o = **radius** of input and output wheels (pulleys)

F_i and F_o = input and output **force** on belt at separate pulleys

D_i and D_o = rim **distance moved** by input and output pulleys

ω_i and ω_o = **angular speed** of input and output pulleys

T_i and T_o = input and output **torque** on separate pulleys

Useful Formulas (No Belt Slippage)

1. $D_i = D_o$ (Distance moved on each rim is the same.)

2. $F_i = F_o$ (Forces are equal and the same as belt tension.)

3. $T_i = F_i \times r_i$ (Input torque equals applied force times lever arm.)

4. $T_o = F_o \times r_o$ (Output torque equals output force times lever arm.)

5. $\omega_i = \dfrac{\theta_i}{t}$ (Angular speed equals angle moved divided by time.)

6. $\omega_o = \dfrac{\theta_o}{t}$ (Angular speed equals angle moved divided by time.)

7. $\text{IMA} = \dfrac{\omega_i}{\omega_o}$ (Ideal mechanical advantage equals ratio of input angular speed to output angular speed.)

8. $\text{IMA} = \dfrac{r_o}{r_i}$ (Ideal mechanical advantage equals the ratio of output wheel to input wheel radii.)

9. $\text{AMA} = \dfrac{T_o}{T_i}$ (Actual mechanical advantage equals the ratio of output torque to input torque.)

10. $\text{IMA} = \text{AMA}$ (This is only true when there's no belt slippage or friction at the shafts.)

Let's examine the effects of belt-drive systems. Let's find the mechanical advantage, torque and speed of a motor-driven compressor (in Example 7-I) where a belt-drive system is used.

Example 7-I: *Air Compressor with a Belt-drive System*

Given: A belt-drive air compressor has a motor pulley 4 inches in diameter. A load pulley is 16 inches in diameter. The motor torque is 4 lb·ft. The motor turns at 480 rpm. Assume no resistance losses.

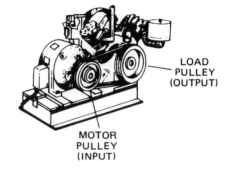

LOAD PULLEY (OUTPUT)

MOTOR PULLEY (INPUT)

Find:
a. The ideal mechanical advantage of the drive system.

b. The torque supplied to the load.

c. The rotational rate of the compressor shaft.

Solution:
a. Use Formula 8 from Table 7-2.

$$IMA = \frac{r_o}{r_i}$$

Substitute values for r_o and r_i, where $r_o = \frac{1}{2}$ (16 in.) = 8 in. for the compressor pulley, and $r_i = \frac{1}{2}$ (4 in.) = 2 in. for the motor pulley.

$$IMA = \frac{8 \text{ in.}}{2 \text{ in.}} = 4$$

b. Use Formulas 9 and 10 from Table 7-2.

Since IMA = AMA, and AMA = $\frac{T_o}{T_i}$, you know that IMA = $\frac{T_o}{T_i}$. Solve for T_o.

$$T_o = IMA \times T_i$$

Substitute in values for IMA and T_i, where $T_i = 4$ lb·ft and IMA = 4.

$$T_o = 4 \times 4 \text{ lb·ft}$$
$$T_o = 16 \text{ lb·ft (load torque)}$$

c. Use Formula 7 from Table 7-2.

$$IMA = \frac{\omega_i}{\omega_o}$$

Solve for ω_o.

$$\omega_o = \frac{\omega_i}{IMA}$$

Substitute in values for ω_i and IMA, where $\omega_i = 480$ rpm and IMA = 4.

$$\omega_o = \frac{480 \text{ rpm}}{4}$$
$$\omega_o = 120 \text{ rpm}$$

Thus the compressor shaft rotates with a speed of 120 rpm.

Let's examine the solution of Example 7-I.

Part a — When you know the diameter of a pulley, you divide by two to obtain the radius. The ratio of the pulley radii (r_o/r_i) is the same as the ratio of the pulley diameters (d_o/d_i). So the ideal mechanical advantage can be found from IMA = r_o/r_i or IMA = d_o/d_i. For Part a of Example 7-I, IMA would be $^8/_2$ (ratio of radii) or $^{16}/_4$ (ratio of diameters). In either case, the answer is "4."

Part b — The output torque was four times greater than the input torque. The output torque was increased. But the output speed was decreased. There's always this kind of trade-off between increased torques and decreased speeds in a drive-belt system.

Part c — The ideal mechanical advantage (IMA) in Part c of Example 7-I is expressed as a ratio of angular speeds of input pulley to output pulley. The larger output pulley developed a higher torque than the input pulley. But it turned at a slower speed.

In the example, the torque increased by a factor of 4, from 4 lb·ft to 16 lb·ft. At the same time, speed decreased by the same factor of 4, from 480 rpm to 120 rpm.

Suppose you reverse the input and output pulleys between the compressor and drive motor. The same compressor described in Example 7-I would work at a higher rpm (1920 rpm), but at a lower torque (1 lb·ft).

In general, a small drive pulley and a large driven pulley give a lower output speed and a higher torque. A large drive pulley and a small driven pulley give a higher output speed and a lower torque.

In Example 7-I, the drive system was said to be free of friction losses and belt slippage. Naturally, this led to a system with an efficiency of 100%. With real belt drives, there's always some loss of energy in the form of heat.

This heat energy loss is due to the flexing of the belts, friction at the pulley shafts, or possible belt slippage in the drive system. In those instances, the actual mechanical advantage is less than the ideal mechanical advantage. Therefore, the efficiency is less than 100%.

WHAT ABOUT GEAR DRIVES?

Two chief differences exist between belt drives and gear drives. First, belt drives cause both pulleys to rotate in the **same** direction. Adjacent gears in gear-drive systems rotate in **opposite** directions.

Second, belt drives sometimes have slippage between belts on pulleys. But gear drives are fitted with gear "teeth." They can't slip. Meshed gears like those in Figure 7-18 don't slip. (Notice how the meshed gears must always turn in opposite directions.)

Table 7-3 sums up the formulas that apply to gear-drive systems.

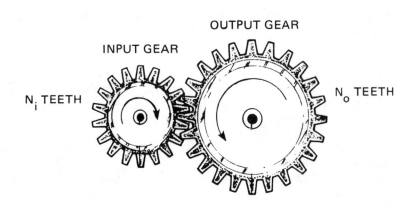

Fig. 7-18 A simple gear drive.

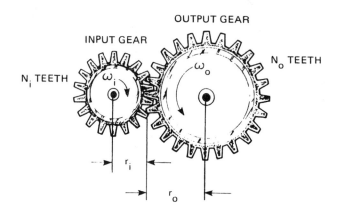

OUTPUT GEAR

INPUT GEAR

N_o TEETH

N_i TEETH

ω_i

ω_o

r_i

r_o

Definition of Symbols

r_i and r_o = **radius** of input and output gears

ω_i and ω_o = **angular speed** of input and output gears

N_i and N_o = total **number** of teeth on input and output gears

Useful Formulas (No Friction)

1. $IMA = \dfrac{r_o}{r_i}$ (IMA equals the ratio of output to input radii.)

2. $IMA = \dfrac{N_o}{N_i}$ (IMA equals the ratio of output to input total teeth count.)

3. $IMA = \dfrac{\omega_i}{\omega_o}$ (IMA equals the ratio of input to output angular speeds.)

4. $IMA = \dfrac{T_o}{T_i}$ (IMA equals the ratio of output to input torques, when friction is not present.)

5. $\dfrac{\omega_i}{\omega_o} = \dfrac{r_o}{r_i}$ (Angular speeds and gear radii are inversely related.)

6. $\dfrac{\omega_i}{\omega_o} = \dfrac{N_o}{N_i}$ (Angular speeds and gear-teeth count are inversely related.)

Example 7-J shows us how to use Table 7-3 to solve a problem that involves gear drives.

Example 7-J: *Final Drive Gear Box on a Bulldozer*

Given: The input gear in the gear box on a bulldozer has 16 teeth. The output gear has 168 teeth. The output gear speed is 28 rpm. This causes the tractor to move forward at 3 mph.

Find:
a. The ideal mechanical advantage of the gear box.

b. The rotational speed of the input gear.

c. The rotational direction of the output gear when the input gear is moving counterclockwise (ccw).

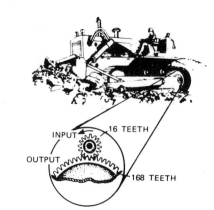

INPUT — 16 TEETH

OUTPUT — 168 TEETH

Solution: a. Use Formula 2 in Table 7-3.

$$IMA = \frac{N_o}{N_i}$$

Substitute values for N_o and N_i, where $N_o = 168$ and $N_i = 16$.

$$IMA = \frac{168 \text{ teeth}}{16 \text{ teeth}}$$

$$IMA = 10.5$$

b. Use Formula 3 in Table 7-3, $IMA = \frac{\omega_i}{\omega_o}$. Solve for ω_i.

$$\omega_i = IMA \times \omega_o$$

Substitute values for IMA and ω_o, where IMA = 10.5 and $\omega_o = 28$ rpm.

$$\omega_i = 10.5 \times 28 \text{ rpm}$$

$$\omega_i = 294 \text{ rpm}$$

c. If the input gear rotates ccw, the output gear rotates in the opposite direction (cw).

WHAT ARE IDLER GEARS?

If the gear drive involves more than two gears on **separate shafts,** the mechanical advantage of one combination is multiplied by the mechanical advantage of the next combination. The product equals the mechanical advantage of the overall system.

If you number the gears in a gear-drive system in sequence, the odd-numbered gears will turn in one direction. The even-numbered gears will turn in the opposite direction. Example 7-K points out these facts.

Example 7-K: *Multiple Gears in a Gear Drive with Separate Shafts* ─────

Given: The gear train shown in the illustration. The gear with 10 teeth is the input gear. Its speed is 20 rpm.

Find: a. The mechanical advantage of the overall system, from input gear to output gear.

b. The direction the output shaft turns if the input shaft turns clockwise (cw).

Solution: a. The mechanical advantage of the system is the mechanical advantage of each combination multiplied by the next combination, and so on.

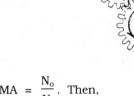

From Formula 2, Table 7-3, $IMA = \frac{N_o}{N_i}$. Then,

$$IMA_1 = \frac{30 \text{ teeth}}{10 \text{ teeth}} \quad \text{(set 1)}$$

$$IMA_2 = \frac{15 \text{ teeth}}{30 \text{ teeth}} \quad \text{(set 2)}$$

$$IMA_3 = \frac{20 \text{ teeth}}{15 \text{ teeth}} \quad \text{(set 3)}$$

The overall IMA is the product of all three:

$$\text{IMA}_{\text{TOT}} = \frac{\cancel{30}}{10} \times \frac{\cancel{15}}{\cancel{30}} \times \frac{20}{\cancel{15}} \qquad \text{(Cancel number of teeth of intermediate gears.)}$$

$$\text{IMA}_{\text{TOT}} = \frac{20}{10} = 2$$

It's clear that only the first gear (N = 10 teeth) and the last gear (N = 20 teeth) contribute to the mechanical advantage. All gears in between add nothing. They transmit torque and change direction. They are called **"idler gears."** Thus, the overall mechanical advantage is always simply:

$$\text{IMA}_{\text{overall}} = \frac{\text{N(output teeth)}}{\text{N(input teeth)}}$$

b. The input gear turns clockwise. Numbering the gears from input to output:

Number 1 (input gear) has 10 teeth and turns clockwise (cw).

Number 2 has 30 teeth and turns counterclockwise (ccw).

Number 3 has 15 teeth and turns clockwise (cw).

Number 4 (output gear) has 20 teeth and turns counterclockwise (ccw).

All odd-numbered gears turn clockwise. Gears number 1 and 3 turn clockwise. All even-numbered gears turn in the opposite direction. So gears number 2 and 4 turn counterclockwise.

WHAT ARE SOME OTHER 'FORCE' TRANSFORMERS IN ROTATIONAL SYSTEMS?

There are other rotational force transformers used in industry. Let's consider a few of the more important ones.

- *Friction Drives* — These devices use two wheels of different shapes similar to gears. But these wheels have no teeth. Torque is transmitted by friction between the rims of the two wheel surfaces. Figure 7-19a shows one example of this type drive—a turntable. An input shaft with different diameters turns a large-diameter turntable wheel. The diameter of the input shaft depends on the desired speed. In this case, the speed at the output is 33 ⅓ rpm, 45 rpm or 78 rpm.

- *Chain Drives* — These devices combine parts of a belt-drive and gear-drive system. The wheels have gear or sprocket teeth cut in them. So there's no slippage. But the chain allows separation of the teeth. So both sprockets turn in the same direction, as do the belt drives. A belt-drive variation of this nonslip principle is the gear belt (or cog belt) where the pulleys and belt have nonslip teeth on them. (See Figures 7-19b, c.)

- *Hybrid Drives* — Many drives have been developed by industry that use the best features of gear, disk and belt drives. They're combined in clever systems and are found on all types of machines—from a typewriter gear drive to a 747 aircraft.

- *Screw Drive* — As mentioned in Subunit 1, an inclined plane can be either a linear or rotational force transformer. Consider an inclined plane curved into an endless circle to form threads on a **screw,** or threads on the common house-leveling jack. An entire fastener industry is built around threaded fasteners or screw drives. Precision metal and woodworking machines move stock or cutting tools by using thread-driven table drives. (See Figures 7-19d, e.)

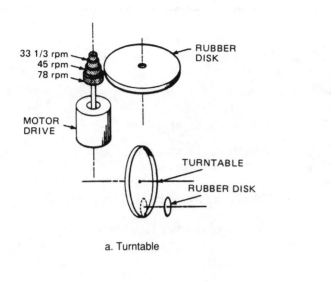

a. Turntable

b. Chain drive

c. Gear belt

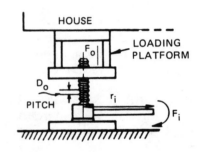

d. House jack

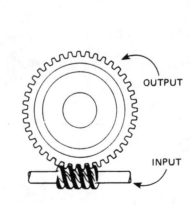

e. Worm drive

Fig. 7-19 Applications in industry.

The following exercises review the main ideas and definitions presented in this subunit, "Force Transformers in Rotational Mechanical Systems." Complete each question.

1. Name three types of rotational force transformers.

2. A garage chain hoist is a variation (or form) of a ____.

3. **Work In** equals **Work Out** in a rotational system. This equality is expressed as:
    a. $(\text{Force} \times \text{Speed})_{input} = (\text{Force} \times \text{Speed})_{output}$
    b. Torque Input = Torque Output
    c. $(\text{Torque} \times \text{Angle Moved Through})_{input} = (\text{Torque} \times \text{Angle Moved Through})_{output}$

4. The ideal mechanical advantage (IMA) of a force transformer is always:
    a. equal to the ratio of Force Out/Force In.
    b. less than the actual mechanical advantage.
    c. equal to or greater than the actual mechanical advantage.
    d. dependent on the presence or absence of friction.

5. The actual mechanical advantage (AMA) of a belt-drive system equals:
    a. the ratio of wheel or pulley radii, r_o/r_i.
    b. the ratio of forces, F_o/F_i.
    c. the ratio of torques, T_o/T_i.
    d. the ideal mechanical advantage, even when friction is present.

6. What's the ideal mechanical advantage (IMA) for a wheel-and-axle force transformer where the input wheel radius is 6 inches and output wheel radius is 2 inches?

7. If the efficiency of a gear drive is 90%, what would the actual mechanical advantage (AMA) be when the ideal mechanical advantage (IMA) is 24?

8. An increase in torque on the output pulley of a belt-and-pulley drive is accompanied by ____ (an increase, a decrease) in angular speed of the output pulley wheel.

9. How does the rotational direction of the third gear in a simple drive train compare to the rotational direction of the fifth gear?

10. Suppose you "cross" the drive belts between two pulleys in a belt-drive transformer, as shown here. What will the direction of the output shaft be compared to the input shaft?

INPUT

OUTPUT

11. What's an idler gear?

12. To get a mechanical advantage of 4 in a gear drive that has a gear with 10 teeth on the input shaft, what size output gear should be chosen?

13. What is the "speed" mechanical advantage for the gear drive in Problem 12 if the torque mechanical advantage is 4?

Math Skills Laboratory

MATH ACTIVITY

Activity: Solving Rotational "Force" Transformer Problems

MATH SKILLS LAB OBJECTIVES

When you complete these activities, you should be able to do the following:

1. **Solve and interpret rotational force transformer problems for mechanical advantage and efficiency.**

2. **Distinguish between force, torque and speed mechanical advantages.**

LEARNING PATH

1. **Read the Math Skills Lab. Give particular attention to the Math Skills Lab Objectives.**

2. **Work the problems.**

ACTIVITY

Solving Rotational 'Force' Transformer Problems

Lathes, bandsaws, drill presses, mills and shapers are machine tools. These tools are more useful when their operating speed can be changed.

A common way to change the operating speed of a rotational mechanical device involves using a "stepped-cone" pulley. This pulley system is shown in Figure 1. The term, "stepped-cone pulley," means that the pulley has a cone shape and has different-diameter pulleys (steps) machined into each complete pulley.

If the pulleys are arranged as shown in Figure 1, the same belt can be connected to corresponding pulley steps. Moving the belt from one pair of pulley steps to another changes the operating speed of the device.

The ratio of circumferences (distance around) of the corresponding pulleys is one thing that determines the speed of the driven shaft. The rpm of the drive pulley also determines the operating speed of the device.

During this lab, you'll learn to find the mechanical advantage of different types of rotational mechanical devices. You'll solve technical problems that involve wheel-and-axle force transformers, belt-drive systems and friction-drive systems.

You should understand ratio and proportion before working these problems. You must also know how to rearrange simple formulas. Also, you should make drawings for each problem. Label forces, torques, radii, angular speeds (and so on) for each problem. This will help you solve the problems correctly.

Problem 1: Given: A stepped-cone drive pulley is mounted on a motor shaft that turns at 1750 rpm (183 rad/sec). The driven stepped-cone pulley is mounted on the driven shaft of a bandsaw. Motor shaft torque is 4 lb·ft under full load. Figure 1 shows the diameter of each step on the pulley. The equations for ideal mechanical advantage are:

$$IMA = \frac{r_o}{r_i} = \frac{T_o}{T_i} = \frac{\omega_i}{\omega_o}$$

The conditions in Problem 1:

(1) drive belt placed on the 4-inch drive pulley and 4-inch driven pulley.

(2) drive belt placed on the 3-inch drive pulley and 6-inch driven pulley.

(3) drive belt placed on the 6-inch drive pulley and 3-inch driven pulley.

Note: Assume ideal efficiency in the system.

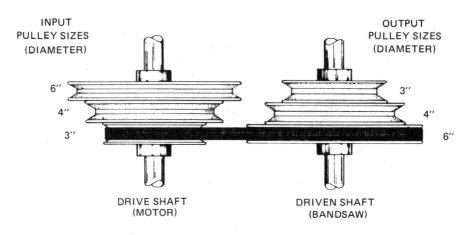

INPUT PULLEY SIZES (DIAMETER)

OUTPUT PULLEY SIZES (DIAMETER)

6″ 4″ 3″ 3″ 4″ 6″

DRIVE SHAFT (MOTOR)

DRIVEN SHAFT (BANDSAW)

Fig. 1 Stepped-cone pulleys.

Find:
   a. The **angular speed** of the bandsaw shaft with conditions (1), (2) and (3) above.

   b. The bandsaw shaft torque with conditions (2) and (3) above.

   c. The **ideal mechanical advantage** of the belt-drive system with conditions (2) and (3) above.

Solution:

Business and industrial buildings have doorways through which bulk materials and equipment can be carried in and out. This doorway is called a "loading dock."

Frequently, the loading dock has a "roll-up" door similar to the one shown in Figure 2. When this type of door opens, it rolls up around a shaft that spans the doorway opening. The shaft around which the door wraps is actually an axle. This axle is driven by a wheel. The wheel is actually a sprocket. And the sprocket is driven by a chain.

Pulling on the chain causes the sprocket to rotate. This causes the axle to rotate. The direction in which the chain is pulled determines whether the door wraps around the axle and opens, or unwraps from the axle and closes.

Problem 2: Given: A sheet-metal roll-up door weighs 400 lb. The chain-operated input wheel (sprocket) is 12 inches in diameter. The shaft (axle) the door cable wraps around is 1.5 inches in diameter. An input force of 50 pounds is required to open or close the door. The mechanical advantage of a wheel-and-axle force transformer can be determined with the following relationships.

$$\text{IMA} = \frac{r_i}{r_o} \quad \text{and} \quad \text{AMA} = \frac{F_o}{F_i}$$

Assume ideal efficiency (no friction) for the door mechanism.

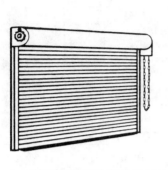

a. Hand-operated chain opener

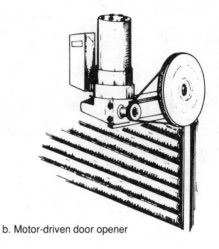

b. Motor-driven door opener

Fig. 2 Roll-up type door.

Find: a. The ideal and actual mechanical advantages of the wheel and axle used with a sheet-metal roll-up door. Are the ideal and actual mechanical advantages equal?

 b. The distance the chain must be pulled to raise a sheet-metal door 5 feet.

Solution:

Problem 3: Given: The sheet-metal door in Problem 2 is replaced with an aluminum roll-up door that weighs 250 pounds.

Find: a. The mechanical advantage needed to raise the aluminum door if the input force remains 50 pounds. Assume ideal efficiency.

 b. The distance the aluminum door will rise when the chain is pulled 5 feet.

Solution:

Problem 4: Given: The conditions in Problem 3.

Find: The minimum wheel (sprocket) diameter that can be used with the aluminum door and provide a mechanical advantage of 5 if the axle diameter remains 1.5 inches.

Note: The ratio of the diameters is the same as the ratio of the radii.

Solution:

When you mix asphalt, you must dry the gravel used in the mix. Excess water must be removed from the gravel. This makes good adhesion between the asphalt and gravel possible.

Figure 3 shows one type of gravel-drying machine. It operates as a friction-drive rotational transformer. The equations that apply are the same as those in Table 7-2.

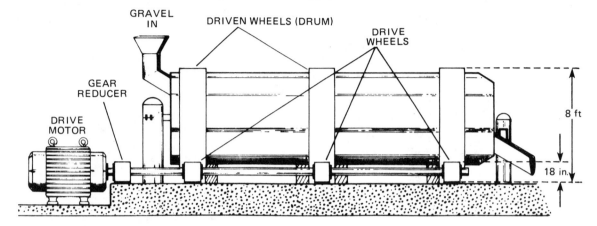

Fig. 3 Gravel dryer.

Problem 5: Given: The dryer drum is 8 feet in diameter. The drum rotates at 99 rpm. The rubber drive wheels are 18 inches in diameter. The drive motor rotates at 1500 rpm. The motor drive is coupled to a gear-type speed reducer. The speed reducer is coupled to the input shaft of the drive wheels.

Find:
    a. The speed (rpm) of the drive wheels.

    b. If the motor operates at 1050 rpm under load, what's the speed reduction (ω_i to ω_o) of the gear reducer?

Solution:

In industry, most electric motors run at shaft speeds of 1050 rpm or 1725 rpm under full load. To use these motor outputs, most gear-train drive systems are made to reduce—rather than increase—speed. Mechanical devices that do this are called "speed reducers." If a speed reducer links directly to the motor housing, it's called a "motorized speed reducer" or a "gear motor."

Problem 6: Given: A gear motor develops 2000 lb·ft of motor torque. It operates with an output shaft speed of 350 rpm. Input motor speed is 1050 rpm.

Find:
    a. The speed reduction ratio of the gear motor.

    b. The torque mechanical advantage of the gear motor.

    c. The output shaft torque.

Solution:

Problem 7: Given: A gear motor has a gear train. The gear train has a 20-tooth gear on the motor drive shaft. It also has a 40-tooth idler gear. There's an 80-tooth gear on the output shaft.

Find:
    a. The speed reduction of the gear motor.

    b. The speed of the motor if the output shaft turns 450 rpm.

    c. The motor torque if output shaft torque is 2000 lb·ft.

Solution:

In industry, machines are more versatile when they can operate at various speeds. You know how stepped-cone pulleys can be used to change the operating speed of a machine. When gears are used to change the speed of a machine, the distance between the driving shaft and the driven shaft is fixed. (It stays the same.)

However, the speed of a gear-driven machine can be changed by installing matched sets of drive and driven gears on the shaft. This is because the gears of these "gear sets" have different diameters. Therefore, different gear sizes can change the shaft speed ratio.

For example, the drive and driven shafts on a machine are to be 4.5 inches apart (center to center). A 3-inch-diameter drive gear and a 6-inch-diameter driven gear would work as one matched set for the machine. A 4-inch-diameter drive gear and a 5-inch-diameter driven gear could be another matched set.

If you add the diameter of each gear in a matched set, you'll notice that the total in this example always equals 9 inches. Why do gears in each matched set have this common total? Because the drive and driven shafts are 4.5 inches apart. And the gears must meet in the middle to mesh.

Problem 8: Given: A matched-gear set on a lathe produces an output shaft speed of 20 rpm when the input shaft speed is 80 rpm. The drive-gear diameter is 4 inches. The driven-gear diameter is 16 inches.

Find:
    a. Ideal mechanical advantage (IMA) of the gear set.

    b. Input torque when the output torque is 20 lb·ft.

    c. Ideal mechanical advantage (IMA) if the input and output gears were exchanged.

    d. Output shaft speed if the drive and driven gears were each 10 inches in diameter.

Solution: (*Hint*: Refer to Table 7-3 for formulas to use.)

Problem 9: Given: Matched-gear sets are used to speed up or slow down feed rollers on a printing press. Two gears are used to drive a roller. A 16-tooth gear is on the input shaft. A 24-tooth gear is on the driven-roller shaft.

Find:
    a. Ideal mechanical advantage of the gear set.

    b. To have the lower output torque on the feed rollers, which gear of a 16-tooth and 24-tooth matched-gear set would you use as the drive gear?

    c. If IMA = 1.6, how many teeth would it take on an input gear if the output gear had 32 teeth?

    d. The direction the roller turns if the input shaft rotates clockwise and there's an idler gear between the drive and driven gears.

Solution: (*Hint*: Refer to Table 7-3 for formulas to use.)

Problem 10: Given: A worm-gear drive is made so that one full turn of the worm gear (input) turns the other gear by one tooth. The pitch of the worm gear is 1 inch.

Find:
    a. Mechanical advantage of this gear system, if the worm gear is meshed with a 26-tooth output gear.

    b. Revolutions made by the output shaft when the worm-gear shaft makes 10 revolutions.

    c. Distance a point on the rim of the output gear moves during the 10 revolutions described in b above.

    d. The output torque, if the input torque on the worm gear is 5 lb·ft.

Solution:

Rotational Mechanical Force Transformers: Gears

LAB OBJECTIVES

When you've finished this lab, you should be able to do the following:

1. *Find the ideal mechanical advantage of a spur-gear drive train. Count the number of teeth and measure the diameters of the various gears.*
2. *Find the ideal mechanical advantage of a worm-and-wheel gear train. Measure the gear-wheel diameter and worm-gear pitch.*
3. *Measure input and output forces in gear trains. Find the actual mechanical advantages of the two drives.*
4. *Find the efficiencies of the two lab gear trains from the measurements of input/output forces and input/output displacements.*

LEARNING PATH

1. ***Preview the lab. This will give you an idea of what's ahead.***
2. ***Read the lab. Give particular attention to the Lab Objectives.***
3. ***Do lab, "Rotational Mechanical Force Transformers: Gears."***

MAIN IDEAS

- *Ideal mechanical advantage of a spur-gear drive train equals the product of all the individual gear ratios.*
- *Ideal mechanical advantage of a worm-and-wheel gear train is equal to the number of teeth on the driven gear.*

A gear is a wheel with notches—called "teeth"—on its rim. Usually, a gear is mounted on a shaft (rod).

Two gears are often positioned so that their teeth mesh, as shown in Figure 1. When one gear turns, its teeth push on the teeth of the other gear. This causes the second gear to move.

When two gears in a mechanical system mesh, one gear drives the other by applying force to it. The gear that applies force is called the "drive"—or "driving" gear. The other gear is called the "driven" gear. The force of the teeth on the driving gear is applied to the teeth of the driven gear. This is what causes it to turn. A driven gear turns its shaft; a driving gear is turned by its shaft.

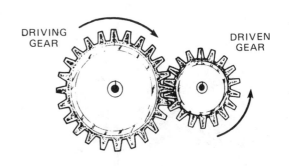

Fig. 1 Drive and driven gears.

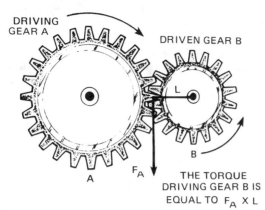

DRIVING GEAR A

DRIVEN GEAR B

L

B

A F_A

THE TORQUE
DRIVING GEAR B IS
EQUAL TO $F_A \times L$

Fig. 2 Torque is applied by one gear to turn the other.

In Figure 2, the force applied by drive-gear A on the driven-gear B, is labeled F_A. The force results from one tooth pushing on another. The lever arm of gear B is equal to the radius of that gear. In Figure 2, the radius of gear B is labeled L.

In Unit 1, *Force*, you learned that torque is the product of force and length of lever arm. In equation form, torque is given as $T = F \times L$. The torque that gear A exerts on gear B (in Figure 2) is equal to the force applied by gear A times the length of the lever arm of gear B.

The equation $T = F \times L$ shows that increasing or decreasing the length of the lever arm increases or decreases the torque produced.

This means that the length of the lever arm also affects the work done by the gears ($W = T \times \theta$), and the rate at which work is done (Power = Work/Time).

Many different types of gears are made. The basic—and perhaps most common—gear is the "spur" gear. (See Figure 3.) A spur gear has a distinguishing feature. Its teeth are parallel with the axis of the gear shaft.

Another common gear is the "bevel" gear. Bevel gears are used to change the axis of the line of rotation along which force is applied. When they're turning, spur gears and bevel gears make lots of noise, even when turning at low speeds. The noise level of a machine can be reduced by using "helical" gears. The teeth of these gears are at an angle to the axis of the gear. Helical gear teeth can be designed to mesh like a spur gear, or to mesh and change the axis of rotation like a bevel gear.

The gears shown in Figure 3 are only a few of the many types of gears that are made. Gears often are used in a machine as a way to move force from one place to another. Gears can also change the rate at which force is applied. Gears are used in many types of machines. Therefore, technicians must understand how size, type and the setup of gears in a machine affect the use and performance of that machine.

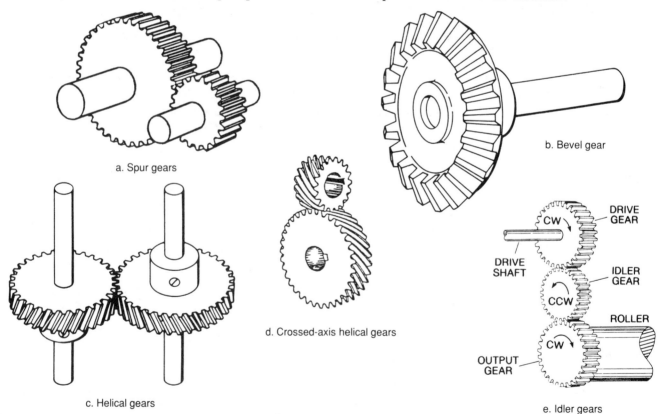

a. Spur gears

b. Bevel gear

c. Helical gears

d. Crossed-axis helical gears

e. Idler gears

DRIVE GEAR

IDLER GEAR

ROLLER

DRIVE SHAFT

OUTPUT GEAR

CW

CCW

CW

Fig. 3 Examples of gears.

EQUIPMENT

Spur-gear train assembly (assembled or unassembled)
Worm-and-wheel gear train
Monofilament line
Small weight set, total 1 kg
Small weight hanger, 50-g type, two
Meterstick/yardstick
Support stand, rods, and clamps

PROCEDURES

Part 1: *Spur-gear Train*

Figure 4 shows a spur-gear train. If your gear train is not yet assembled, refer to this figure. Follow the instructions of Steps 2 through 6. If your gear train is already assembled, begin with Step 1. Then skip to Step 7 and proceed.

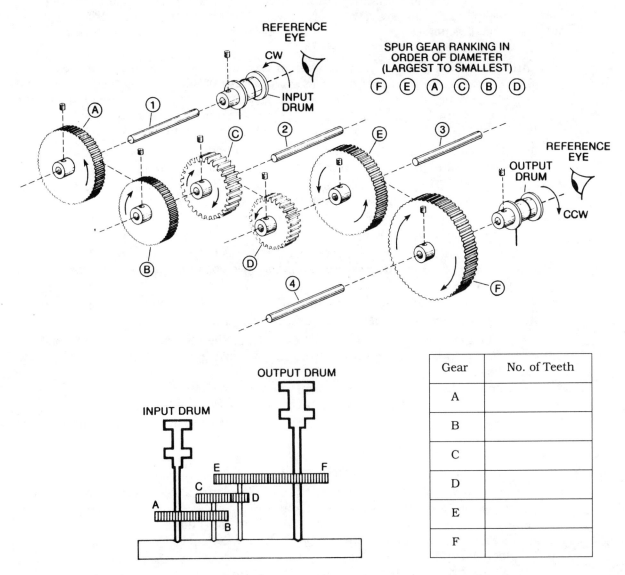

Gear	No. of Teeth
A	
B	
C	
D	
E	
F	

Fig. 4 Lab setup for spur-gear train.

1. Count the number of teeth on each gear. Record in a table, like the one shown in Figure 4.

2. On shaft #1, mount gear A and the input drum. Secure both the gear and the drum to the shaft so that they turn together.

3. On shaft #2, mount gears B and C. Secure both gears to the shaft so that these gears turn together.

4. Assemble shaft #3, and mount gears D and E in the same way.

5. Assemble shaft #4 and gear F and the output drum, as in Step 2.

6. Adjust the position and alignment of each gear shaft so that you can get the following mesh pattern (see side view of gears in Figure 4):
   Gear A meshes with gear B.
   Gear C meshes with gear D.
   Gear E meshes with gear F.

7. From the material you've studied about gears and your look at the lab setup, write your predictions for the following items, with corresponding letters, on your data sheet.
   a. The ideal mechanical advantage (IMA) of the gear train.
   b. The approximate amount of input force it takes to lift a total load of 200-gm weight at the output. (This output weight will be called the "load.")
   c. The direction the monofilament line on the output drum will move with respect to line on the input drum.

8. Attach an 8-inch length of monofilament line to each of the two drums. Wrap the line connected to the input drum in a **clockwise** direction. Use "eye" in Figure 4 as a reference. Also wrap the output drum in a **clockwise** direction, leaving 6 inches of line hanging down. See "eye" for reference. Be sure to secure one end of the line to the drum so the line doesn't slip on the drum.

9. While holding the gears so that they don't turn—connect weight hangers (of equal weight) to each line. Place a 200-gm weight on the output weight hanger.

10. Measure the distance from the weight hanger to the drum on the input and the output. Record these values on your data sheet, as follows:
    Output drum to weight hanger distance (d_o initial) = ____
    Input drum to weight hanger distance (d_i initial) = ____

11. Place a 20-gm weight on the input weight hanger. Increase the input weight in small steps of 10 or 20 gm until the load moves **upward** at a slow and constant speed.

12. When either the input or output lines is near its end of travel, stop the gear train. To stop it, place your finger on any one of the gears.

13. With the gears stopped, as in Step 12, measure the distance from drum weight to hanger for both input and output drums. Record the values on your data sheet, as follows:
    Output drum to weight hanger distance (d_o final) = ____
    Input drum weight to weight hanger distance (d_i final) = ____

14. Also record the actual weight added to the weight hanger at the input and output ends:
    Input (weight) = ____
    Output (load) = 200-gm weight

15. Remove all weights from weight hangers. Place the 200-gm weight on the input drum weight hanger, leaving the weight hanger on the output end empty. Once more, make your predictions on your data sheet, as follows:
    IMA of reversed drive train = ____
    Force needed to lift 200-gm load = ____
    Output rotation with reference to **input rotation**: ____ (same or opposite)

16. Once more, add weights in small steps of 10 or 20 gm to the empty weight hanger—at gear F in Figure 4—until the 200-gm weight is lifted at a slow and steady speed. Record as follows:

Actual input force (weights added) = ____

Output force (load) = 200-gm weight

Part 2: *Worm-and-wheel Gear Train*

Figure 5 shows the proper setup for this portion of the lab. Rotate the input drum and observe the direction of the output drum.

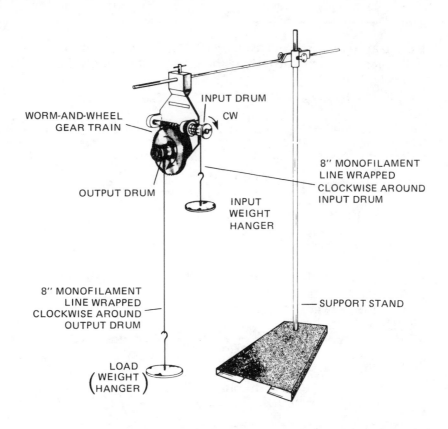

WORM-AND-WHEEL
GEAR TRAIN

INPUT DRUM

CW

8" MONOFILAMENT
LINE WRAPPED
CLOCKWISE AROUND
INPUT DRUM

OUTPUT DRUM

INPUT
WEIGHT
HANGER

8" MONOFILAMENT
LINE WRAPPED
CLOCKWISE AROUND
OUTPUT DRUM

SUPPORT STAND

LOAD
(WEIGHT
HANGER)

Fig. 5 Lab setup for worm-and-wheel gear train.

1. Attach monofilament lines on both the input and load (or output) drums. Wrap the lines so that a downward motion of line on the input drum results in an upward motion of line at the output drum. Each line should be 8 inches or more long. Attach empty weight hangers.

2. Measure the initial position of each weight hanger with respect to its drum. Record these measurements under Part 2 on your data sheet, as follows:

Distance between input drum and weight hanger (d_i initial) = ____

Distance between output drum and weight hanger (d_o initial) = ____

3. Place a 500-gm weight on load weight hanger. Record weight of load:

Load = 500-gm weight

4. Place a 20-gm weight on the input weight hanger. Increase by steps of 10 or 20 gm until the load begins to move upward slowly at a constant speed.

5. Record the weight added to the input weight hanger to lift the load, as described in Step 4:

Input Force = ____

6. Measure the final position on each weight hanger with respect to its drum. Record these measurements on your data sheet:
   Distance between input drum and weight hanger (d_i final) = ____
   Distance between output drum and weight hanger (d_o final) = ____

7. Do not change the final position of the weights from Step 6. Remove about half of the weight from the input weight hanger. Does the worm-and-wheel gear train run backward compared to the previous steps?

8. Count the number of teeth on the driven gear. Record on your data sheet:
   N_o = ____

Calculations

1. Find the ideal mechanical advantage (IMA) of each gear train.
   a. IMA for the spur-gear train is the product of the individual gear ratios involved.
      1) In the "forward" direction:
      $$\text{IMA} = \left(\frac{N_B}{N_A}\right)\left(\frac{N_D}{N_C}\right)\left(\frac{N_F}{N_E}\right) = \underline{\quad} \quad \text{(See \textbf{Part 1}, Steps 1 through 14.)}$$
      2) In the "reversed" direction:
      $$\text{IMA} = \left(\frac{N_E}{N_F}\right)\left(\frac{N_C}{N_D}\right)\left(\frac{N_A}{N_B}\right) = \underline{\quad} \quad \text{(See \textbf{Part 1}, Steps 15 and 16.)}$$
   b. IMA for the worm-and-wheel gear train.
      $$\text{IMA} = \frac{\text{Number of Teeth on Driven Gear}}{1} = \frac{N_o}{1} = N_o \quad \text{(See \textbf{Part 2}, Step 8.)}$$

2. Find the actual mechanical advantage (AMA) of each gear train.
   a. AMA for the spur-gear train.
      1) In the "forward" direction:
      $$\text{AMA} = \frac{\text{Output (load)}}{\text{Input (weight)}} = \underline{\quad} \quad \text{(See \textbf{Part 1}, Step 14 for data.)}$$
      2) In the "reversed" direction:
      $$\text{AMA (rev)} = \frac{\text{Output (load)}}{\text{Input (weight)}} = \underline{\quad} \quad \text{(See \textbf{Part 1}, Step 16 for data.)}$$
   b. AMA for the worm-and-wheel gear train:
      $$\text{AMA} = \frac{\text{Load}}{\text{Input Force}} = \underline{\quad} \quad \text{(See \textbf{Part 2}, Steps 3 and 5 for data.)}$$

3. Find the efficiency of each gear train by the basic relationship of:
   $$\text{Eff} = \frac{\text{Work Out}}{\text{Work In}} \times 100\%$$
   a. Find the displacement for each gear train.
      1) Displacement for the spur-gear train (see **Part 1**, Steps 10 and 13):
      $D_i = (d_i \text{ final}) - (d_i \text{ initial}) = (\underline{\quad}) - (\underline{\quad}) = \underline{\quad}$
      $D_o = (d_o \text{ initial}) - (d_o \text{ final}) = (\underline{\quad}) - (\underline{\quad}) = \underline{\quad}$
      2) Displacement for the worm-and-wheel gear train (see **Part 2**, Steps 2 and 6):
      $D_i = (d_i \text{ final}) - (d_i \text{ initial}) = (\underline{\quad}) - (\underline{\quad}) = \underline{\quad}$
      $D_o = (d_o \text{ initial}) - (d_o \text{ final}) = (\underline{\quad}) - (\underline{\quad}) = \underline{\quad}$

b. Find Work In for each gear train.
   1) For the spur-gear train (**Part 1,** Step 14 and Calculation 3.a.1):

   $$W_i = \text{Input (Weight)} \times D_i = (\underline{\quad}) \times (\underline{\quad}) = \underline{\quad}$$

   2) For the worm-and-wheel gear train (**Part 2,** Step 5 and Calculation 3.a.2):

   $$W_i = \text{Input Force} \times D_i = (\underline{\quad}) \times (\underline{\quad}) = \underline{\quad}$$

c. Find Work Out for each gear train.
   1) For the spur-gear train (**Part 1,** Step 14 and Calculation 3.a.1):

   $$W_o = \text{Output (load)} \times D_o = (\underline{\quad}) \times (\underline{\quad}) = \underline{\quad}$$

   2) For the worm-and-wheel gear train (**Part 2,** Step 3, and Calculation 3.a.2):

   $$W_o = \text{Load} \times D_o = (\underline{\quad}) \times (\underline{\quad}) = \underline{\quad}$$

d. Find the efficiency of each gear train.
   1) For the spur-gear train (Calculations 3.b.1 and 3.c.1):

   $$\text{Eff} = \frac{W_o}{W_i} \times 100\% = \frac{(\underline{\quad})}{(\underline{\quad})} \times 100\% = \underline{\quad}$$

   2) For the worm-and-wheel gear train (Calculations 3.b.2 and 3.c.2):

   $$\text{Eff} = \frac{W_o}{W_i} \times 100\% = \frac{(\underline{\quad})}{(\underline{\quad})} \times 100\% = \underline{\quad}$$

WRAP-UP

1. Is the action one-way or reversible for the
   a. spur-gear train?
   b. worm-and-wheel gear train?

2. Were the predicted IMAs of the spur-gear and the worm-and-wheel gear trains close to the AMAs?

3. Gear trains can be used to change the orientation of input to output shaft rotation. Does the spur-gear train do this? Under what conditions? Does the worm-and-wheel gear do this?

4. What steps could you take to improve the efficiency of each of these gear trains?

Force Transformers: Belts and Pulleys

LAB OBJECTIVES

When you've finished this lab, you should be able to do the following:

1. **Find the mechanical advantage and efficiency of a belt-drive system.**
2. **Identify the pros and cons of V-belt and cogged-belt drive systems.**

LEARNING PATH

1. **Preview the lab. This will give you an idea of what's ahead.**
2. **Read the lab. Give particular attention to the Lab Objectives.**
3. **Do lab, "Force Transformers: Belts and Pulleys."**

MAIN IDEAS

- **A belt-and-pulley system is a common force transformer.**
- **Belt-and-pulley systems can be used to increase or decrease the operating speed of a machine.**
- **A belt-and-pulley system can be used to increase or decrease the mechanical advantage of a system.**

Rotational mechanical systems often use belt-drive devices. There are three common types of belts. These are the "flat belt," the "V-belt" and the "cog belt."

In addition to these three types, there's a hybrid belt—the chain. ("Hybrid" means a mixture of types.) A common use of each type is shown in Figure 1.

Fig. 1 Examples of belt drives.

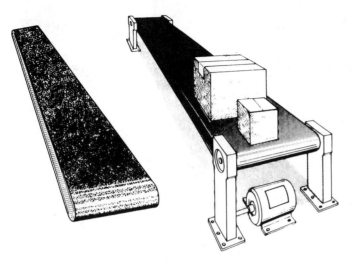

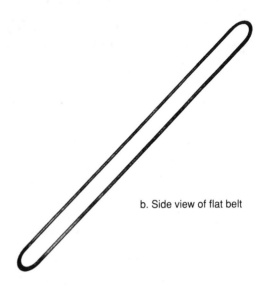

b. Side view of flat belt

a. Flat belts are usually long and wide.

Fig. 2 Flat belt.

A flat belt looks a lot like its name. It's flat. (See Figure 2.) It's long and wide. The flat belt usually is used on machines that work at a slow speed. Some work requires a long, wide belt. For example, a conveyor belt is a "flat belt."

Fig. 3 Cross section of a V-belt.

V-belts are used when the belt must move fast. Cars use V-belts to drive the water pump. V-belts also power the steering pump, alternator, air-conditioning compressors and emission-control devices. Figure 3 shows a cross-sectional view of a V-belt. In this view, the belt has a shape similar to the letter "V." That's why it's called a "V-belt."

Both V-belts and flat belts tend to slip on the pulleys that drive or are driven by them. In some applications, the rotation of two pulleys

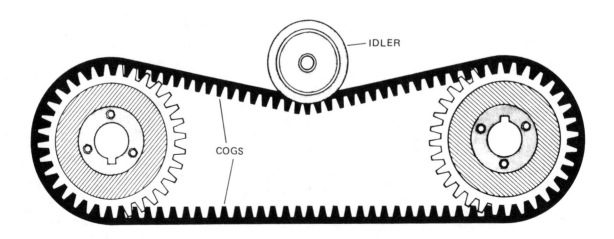

Fig. 4 Side view of a cog belt.

must be "timed"—or well-synchronized. (That is, the rotations must occur at the same time, or in harmony, with another.) A "cog belt" is used when this is important.

The cog belt has projections, known as "cogs," on the side of the belt. This is shown in Figure 4. Cogs are evenly spaced around the belt. Pulleys used with cog belts have notches around the rim. Cogs on the belt mesh with the notches on the pulley. This keeps the belt from slipping on the pulley. Compared to V-belts and flat belts, cog belts can work at high speeds and can transmit large amounts of force.

Chains are hybrid belts. Basically, they're a gear and a belt. (See Figure 5.) The gear, called a "sprocket," has teeth that mesh with holes in the belt (chain). This action prevents the chain from slipping on the sprocket. Because of that, chains are also used when the angular motion of two sprockets must be timed. Chains can transmit lots of force compared to cog belts. However, chains are best suited for low-speed work. Special high-speed chains are made, but

they're very expensive compared to the "normal" type of chain or cog belts.

Almost every machine with moving parts uses either a belt-drive system or a gear-drive system to move those parts. Belt drives are often used with many types of machines. Because of that, technicians often must find out how a belt-drive system affects the work of other systems on a machine.

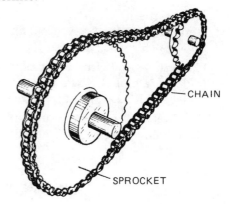

Fig. 5 Chain and sprockets.

LABORATORY

EQUIPMENT

Stepped belt-drive assembly, with cogged motor pulley equipped with safety enclosure
DC permanent-magnet motor, nominally 0-12 V
DC power supply, 20 V, 10 A
Cogged pulley attachment for DC motor
Stroboscope
C-clamp, two
Vernier calipers
Ruler (12 in., with centimeter scale)

PROCEDURES

Part 1: Stepped Belt-drive Assembly

1. Set up the pulley assembly, as shown in Figure 6. Use C-clamps to fasten the motor-and-pulley assembly to rigid supports. The motor-and-pulley assembly should be far enough apart to give it good belt tension.

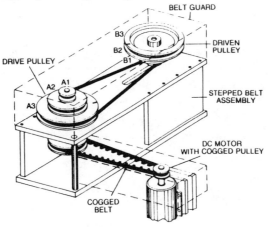

Fig. 6 Lab setup for stepped belt-drive assembly.

Subunit 2: Force Transformers in Rotational Mechanical Systems 73

2. Measure the diameter of each of the pulleys and its steps. Then record the values on your data sheet as follows:

Motor-pulley Diameter (d_m) = ____
Drive Pulley (smallest to largest) d_{A1} = ____
d_{A2} = ____
d_{A3} = ____
Driven Pulley (smallest to largest) d_{B1} = ____
d_{B2} = ____
d_{B3} = ____

3. Loosen the driven step-pulley retaining bolts. Slide that pulley toward the *drive* step-pulley. Install the V-belt on the largest drive pulley and smallest driven pulley. This will be called "Position 1."

Place the V-belt under tension by sliding the driven pulley away from the drive pulley. Adjust the belt tension by placing a long straightedge (meterstick) on the pulleys. This is shown in Figure 7.

The belt is too lose if it can be pushed more than $\frac{3}{8}$ inch away from the straightedge. The belt is too tight if it can't be pushed at least $\frac{1}{4}$ inch away from the straightedge.

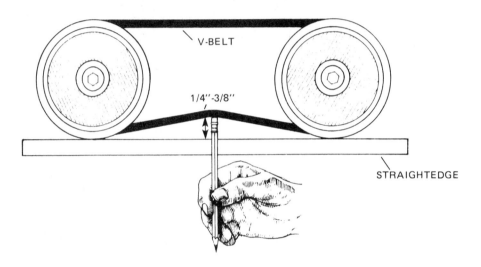

Fig. 7 Adjust belt tension so that belt deflects $\frac{1}{4}$" - $\frac{3}{8}$" between the straightedge and the belt.

4. With the safety enclosure in place, connect the DC motor to the power supply. Be sure the power supply is turned OFF. The **voltage control should be set on "minimum."** Plug in the power supply to the wall outlet. Then turn the power supply ON. Adjust the output voltage to 6 volts. This will cause the motor to turn the pulleys. **CAUTION: Keep fingers and loose objects away from moving belts and pulleys.**

5. Use the stroboscope to measure three things. First, measure the rpm of the motor. Then measure the rpm of the step-drive pulley. Then measure the rpm of the step-driven pulley. Record these speeds in Data Table 1 in the correct columns and row, labeled "Position 1," "6 volts."

6. Increase the power supply output voltage to 10 volts. Repeat Step 4. Record your data in the correct columns and row, labeled "Position 1," "10 volts."

7. Turn the power supply OFF. Return the voltage control to "minimum." Move the V-belt to the smallest step of the drive pulley and largest step of the driven pulley. Readjust the belt tension. This will be called "Position 2."

DATA TABLE 1

Position Number	Volts Across Motor	Motor rpm	Drive Wheel rpm	Driven Wheel rpm
1	6 volts 10 volts			
2	6 volts 10 volts			
3	6 volts			

8. Repeat Steps 4 and 5. Record the rpm measurements in the correct columns and row, labeled "Position 2," "6 volts," and "10 volts," respectively.

9. Turn the power supply OFF. Turn the voltage control to "minimum." Move the V-belt to the middle step of the drive and driven pulleys. Readjust the belt tension. This will be called "Position 3."

10. Turn the power supply ON. Adjust the voltage level to 6 volts. Use the strobe to measure the rpm of each pulley. Record the rpm measurements in the correct columns of the row, labeled "Position 3," "6 volts." On your data sheet, note the direction of rotation of the drive pulley and driven pulley. Is the direction of rotation clockwise (cw) or counterclockwise (ccw)?

11. Adjust the V-belt tension to allow one inch of slack. Turn the power supply ON. Use the strobe to measure the rpm of each pulley. Record separately on your data sheet. Again, note the direction of rotation of the drive and driven pulleys.

   To keep the loosened belt from jumping off the pulley, have a lab partner hold a pencil or piece of tubing against the largest rim of the driven pulley. Use the strobe to measure the rpm of the driven pulley. Record this measurement on a separate sheet of paper. Turn the power OFF.

Part 2: *Cog-belt Drive*

1. Notice that a cog belt is used to transfer energy from the motor to the stepped-drive assembly, as shown in Figure 8.

2. Vary the tension slightly between the motor drive pulley and driven pulley. Does the cog belt slip on either of the pulleys?

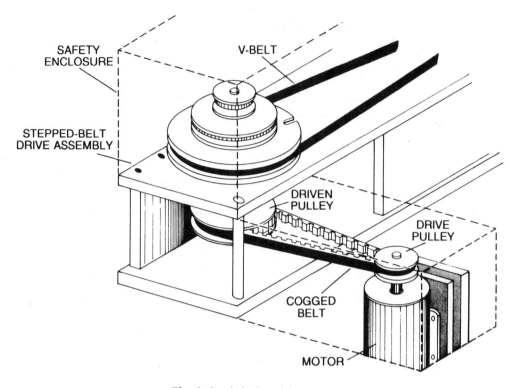

Fig. 8 Cog-belt-drive lab setup.

WRAP-UP

Use Data Table 1 and your data sheet notes to help answer these questions:

1. Can stepped pulleys be used to change the output speed of a device with respect to its input speed? Explain your answer.

2. How did changing the speed of the motor affect the speed of the drive and driven pulleys?

3. Find the ideal mechanical advantage of the drive system for:
      Position 1: IMA = ____
      Position 2: IMA = ____

4. What happened to the output torque and speed (compared to the input) when the belt was running in the middle step of both the drive and driven pulleys?

5. Explain the advantage of using a cog belt rather than a V-belt to drive a device.

6. Compare the IMA of the drive system using pulley diameters with the IMA using rotational speeds. Do they agree? What could cause them to disagree?

Student Challenge ———————————————————————

Assuming no belt slippage and a motor speed of 600 rpm, what are the fastest and the slowest rpm possible for the driven step-pulley on your equipment?

Review

1. *View and discuss the video, "Force Transformers in Rotational Mechanical Systems."*

2. *Review the Objectives and Main Ideas of the print materials in this subunit.*

3. *Your teacher may give you a test over Subunit 2, "Force Transformers in Rotational Mechanical Systems."*

SUBUNIT 3

Force Transformers In Fluid Systems

SUBUNIT OBJECTIVES

When you've finished reading this subunit and viewing the video, "Force Transformers in Fluid Systems," you should be able to do the following:

1. *Relate input work to output work for a hydraulic jack.*
2. *Find the mechanical advantage of a hydraulic jack.*
3. *Explain how pressure is amplified in a pressure intensifier.*
4. *Find the efficiency of a pressure intensifier.*
5. *List or identify various fluid transformers.*
6. *Measure the mechanical advantage of a fluid transformer.*
7. *Identify workplace applications where technicians use fluid transformers.*

LEARNING PATH

1. *Read this subunit, "Force Transformers in Fluid Systems." Give particular attention to the Subunit Objectives.*
2. *View and discuss the video, "Force Transformers in Fluid Systems."*
3. *Participate in class discussions.*
4. *Watch a demonstration about fluid force transformers.*
5. *Complete the Student Exercises.*

MAIN IDEAS

- *A hydraulic jack (or lift) is a mechanical device that uses a fluid medium to amplify force.*
- *In a hydraulic jack or lift, output force is greater than input force. But load movement at the output end is less than force movement at the input end.*
- *A pressure intensifier uses fluids to amplify pressure.*
- *In a pressure intensifier, output pressure is greater than input pressure. But volume of fluid moved at the input end is greater than volume of fluid moved at the output end.*
- *Hydraulic lifts, hydraulic jacks and pressure boosters are important applications of force transformers.*

Hydraulic jacks and pressure boosters are examples of force transformers in fluid systems. The hydraulic jack is actually a mechanical force transformer, since it amplifies force. However, the hydraulic jack does this by using a fluid under pressure. So we're going to talk about how it works as a fluid force transformer.

The pressure booster—or pressure intensifier—is a true forcelike transformer in fluid systems. That's because the pressure booster amplifies pressure (not force). And it's *pressure,* not force, that is the *forcelike* quantity in fluid systems, as you already know.

Fluid transformers can use either a gas (pneumatic) or a liquid (hydraulic) as the pressurized fluid. Pneumatic fluid transformers have an efficiency that's far below their ideal efficiency. Hydraulic fluid transformers, on the other hand, come close to their ideal efficiency.

The difference comes about because gases are compressible. Gases are affected more by changes in pressure and temperature than are liquids.

Hydraulic transformers are easier to understand than pneumatic fluid transformers—yet they still show the idea of "force" transformers clearly. So, we're going to discuss only hydraulic-type force transformers in this subunit.

WHAT'S THE RELATIONSHIP BETWEEN INPUT WORK AND OUTPUT WORK IN A FLUID FORCE TRANSFORMER?

We've already said a force transformer is a coupling device that has work done on it by an input force. In turn, the force transformer does work on a load by amplifying the output force or displacement.

We described input work in linear mechanical systems by the general equation for work:

$$\text{Work In} = F_i \times D_i$$

Here, F_i is the *input force* applied. D_i is the distance through which the *input force* moves.

In a fluid transformer—or a hydraulic transformer for our purposes here—input work also equals the pressure applied (p_i) times the volume displacement (ΔV_i) at the input end.

$$\text{Work In} = p_i \times \Delta V_i$$

The input work done on a fluid transformer at the input end is changed to output work on the load at the output end. If the transformer is 100% efficient (ideal efficiency), **work output** equals **work input.**

For fluid transformers, this becomes:

$$\text{Work Input} = \text{Work Output}$$

$$p_i \times \Delta V_i = p_o \times \Delta V_o$$

where: p_i = input pressure
  p_o = output pressure
  ΔV_i = volume moved at input end
  ΔV_o = volume moved at output end

HOW DOES A HYDRAULIC JACK AMPLIFY FORCE?

Figure 7-20 shows a hydraulic jack and a line drawing of the jack's important parts. Simply put, a hydraulic jack amplifies an input force (F_i) into a much larger output force (F_o).

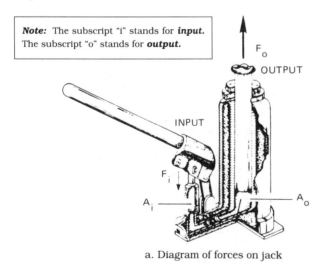

Note: The subscript "i" stands for *input.* The subscript "o" stands for *output.*

a. Diagram of forces on jack

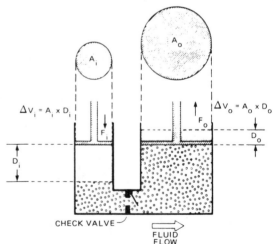

b. Diagram of fluid in jack

Fig. 7-20 Hydraulic jack.

For example, if the mechanical advantage is 5, a 100-pound force applied on the input piston becomes 500 pounds at the load end.

(Don't forget, though, that the handle on the jack is itself a **second-class lever.** So it takes the force applied at the input end of the handle and amplifies it to produce a larger force pressing down on the input piston. The overall mechanical advantage of the jack involves both that of the handle and of the hydraulic pistons.)

Figure 7-20a shows the input piston force (F_i) and cylinder cross section (A_i) at the input end of the jack. It also shows the output force (F_o) and cylinder cross section (A_o) at the output end.

Figure 7-20b shows the input parameters F_i (input force), A_i (input cross section), D_i (input piston movement) and ΔV_i (input volume displaced). The corresponding terms—F_o, A_o, D_o and ΔV_o—are shown at the output end.

In an incompressible fluid (one that can't be squeezed) like water or oil, the pressure is the same everywhere inside the jack. Therefore, $p_i = p_o$.

Because the fluid can't be squeezed or compressed, whatever volume of fluid is moved at the input end (ΔV_i) must also be moved at the output end (ΔV_o). Therefore, $\Delta V_i = \Delta V_o$. This makes input work equal to output work:

$$p_i \times \Delta V_i = p_o \times \Delta V_o$$

since $p_i = p_o$ and $\Delta V_i = \Delta V_o$.

Remember, you can't get more work out than you put in. If there's some resistance in the force transformer, you'll get less work out than in.

As you know, when force is amplified, it's done at the expense of movement. That's just as true in the hydraulic jack as it is in the block and tackle.

For example, if $F_o = 10\ F_i$ (in Figure 7-20), a mechanical advantage of 10, then $D_o = \frac{1}{10}\ D_i$. Thus, if the input force (F_i) moves 30 cm, the output force (F_o) will lift the load only 3 cm.

How does that happen? It's related to the fact just mentioned—that liquids are incompressible. Liquids can't be squeezed into smaller volumes. Liquids can be moved, or displaced, but they can't be compressed.

Look at Figure 7-21. The input cylinder is shown on the left. The output cylinder is shown on the right.

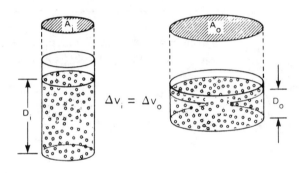

Fig. 7-21 The same fluid in two different cylinders.

The input piston, of cross-sectional area A_i, moves down the cylinder a distance D_i. It displaces a volume ΔV_i. This same volume must be displaced at the output end. Thus, the output piston, of cross-sectional area A_o, moves up the cylinder (while lifting the load) a distance D_o. In doing so, it displaces a volume ΔV_o that's equal to ΔV_i.

Student Challenge ─────────────────────────────

If you want to test your math skills, see if you can follow the logic given here.

$\Delta V_i = \Delta V_o$ (Since a liquid can't be compressed, a volume change at the input must equal that at the output.)

$A_i \times D_i = A_o \times D_o$ (The volume of the "cylinder of liquid moved" equals the area of the cross section times the piston distance moved.)

$\dfrac{\cancel{A_i} \times D_i}{A_i D_o} = \dfrac{A_o \times \cancel{D_o}}{A_i D_o}$ (Divide each side of the equation above by "$A_i D_o$" and cancel like units.)

$\dfrac{D_i}{D_o} = \dfrac{A_o}{A_i}$ (The ratio of distances [input to output] equals the **inverse** ratio of areas [output to input].)

Now let's consider the work equation. Remember, when there's no resistance, **work in** equals **work out**.

$$F_i \times D_i = F_o \times D_o$$ (Work is defined as force times distance.)

$$\frac{\cancel{F_i} \times D_i}{\cancel{F_i} D_o} = \frac{F_o \times \cancel{D_o}}{F_i \cancel{D_o}}$$ (Divide each side by "$F_i D_o$" and cancel like terms.)

$$\frac{F_o}{F_i} = \frac{D_i}{D_o}$$ (The ratio of forces [output to input] equals the **inverse** ratio of distances moved [input to output].)

If there's no friction to worry about, the mechanical advantage for the jack is given by $\frac{F_o}{F_i}$. Since $\frac{F_o}{F_i} = \frac{D_i}{D_o}$, and $\frac{D_i}{D_o} = \frac{A_o}{A_i}$ as just shown, you can write three equivalent equations for the mechanical advantage of the jack:

$$IMA = \frac{F_o}{F_i} \qquad IMA = \frac{D_i}{D_o} \qquad IMA = \frac{A_o}{A_i}$$

Any of these three equations can be used to find the mechanical advantage of a jack.

Table 7-4 sums up the important relationships for a hydraulic jack. (Notice that these apply to the hydraulic part alone and not the handle.)

TABLE 7-4: FORCE TRANSFORMER FORMULAS FOR A HYDRAULIC JACK

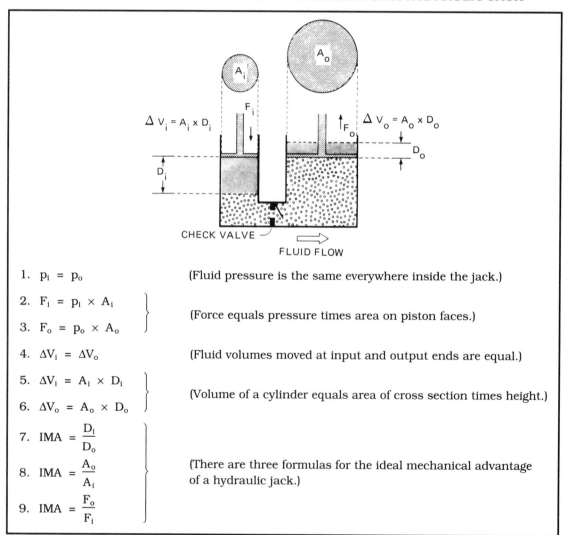

1. $p_i = p_o$ (Fluid pressure is the same everywhere inside the jack.)

2. $F_i = p_i \times A_i$

3. $F_o = p_o \times A_o$ (Force equals pressure times area on piston faces.)

4. $\Delta V_i = \Delta V_o$ (Fluid volumes moved at input and output ends are equal.)

5. $\Delta V_i = A_i \times D_i$

6. $\Delta V_o = A_o \times D_o$ (Volume of a cylinder equals area of cross section times height.)

7. $IMA = \frac{D_i}{D_o}$

8. $IMA = \frac{A_o}{A_i}$ (There are three formulas for the ideal mechanical advantage of a hydraulic jack.)

9. $IMA = \frac{F_o}{F_i}$

Table 7-4 lists the formulas you need to work problems that involve hydraulic jacks—or hydraulic lifts. Now try several of these formulas to be sure you understand what's going on.

Suppose that $A_o = 8$ in^2 and $A_i = 2$ in^2 for the hydraulic jack shown in Figure 7-20. If the input force F_i is 50 lb, what's the output force? Solve the problem this way. From Table 7-4, Formula 8:

$$IMA = \frac{A_o}{A_i} = \frac{8 \text{ in}^2}{2 \text{ in}^2} = 4 \qquad \text{(The ideal mechanical advantage is 4.)}$$

From Table 7-4, Formula 9:

$$IMA = \frac{F_o}{F_i} \qquad \text{where:} \quad F_i = 50 \text{ lb}$$
$$IMA = 4$$

Rearrange the formula and substitute values to get:

$F_o = F_i \times IMA$
$F_o = 50 \text{ lb} \times 4$
$F_o = \cdot 200 \text{ lb}$ (This is the output force.)

The hydraulic jack has amplified the force four times, from 50 lb to 200 lb.

If the piston face at the input end moves down 4 inches (D_i), how far (D_o) does the load move up? You probably guessed "1 inch"—and that's correct. But here's the way to prove it. From Table 7-4, Formula 7:

$$IMA = \frac{D_i}{D_o} \qquad \text{where:} \quad D_i = 4 \text{ in.}$$
$$IMA = 4$$

Rearrange the formula and substitute values to get:

$$D_o = \frac{D_i}{IMA} = \frac{4 \text{ in.}}{4} = 1 \text{ in.}$$

So when the 50-pound input force moves down 4 inches, the 200-pound output force moves up only 1 inch. Again, you see the usual *force transformer* trade-off. If the output force is amplified x-times the displacement is reduced x-times.

Example 7-L shows the mechanical advantage calculation for a hydraulic jack. Notice that the same volume of fluid moves from the input cylinder to the output cylinder. The input force is a linear mechanical force and the output force or load force is also a linear mechanical force.

Therefore, a hydraulic jack is a *linear mechanical force transformer* that uses fluid as the mechanical coupling.

Example 7-L: *Mechanical Advantage of a Hydraulic Jack* ————————

Given: The hydraulic jack shown in the diagram with the following known values:

        Input piston face area (A_i) = 1 in^2
        Output piston face area (A_o) = 4 in^2
        Input piston travel (D_i) = 4 in.
        Load (F_o) = 200 lb

Find: a. The ideal mechanical advantage of the jack.

          b. The input force of the input piston.

          c. The distance the load moves for each handle stroke.

          d. Volume of fluid transferred.

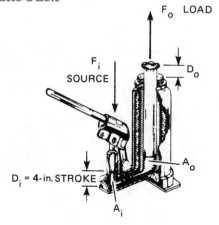

Solution: a. Use Formula 8 in Table 7-4.

$$IMA = \frac{A_o}{A_i} \qquad \text{where: } A_o = 4 \text{ in}^2$$
$$A_i = 1 \text{ in}^2$$

$$IMA = \frac{4 \text{ in}^2}{1 \text{ in}^2} = 4 \qquad \text{(The ideal mechanical advantage is 4.)}$$

b. Since IMA = 4 and the load is 200 lb, input force must be 50 lb. But you can use Formula 9, Table 7-4, to prove your answer.

$$IMA = \frac{F_o}{F_i}$$

$$F_i = \frac{F_o}{IMA} \qquad \text{(Rearrange equation.)}$$

Substitute in known values: IMA = 4 and F_o = 200 lb.

$$F_i = \frac{200 \text{ lb}}{4} = 50 \text{ lb}$$

c. Use Formula 7, Table 7-4, to find the distance the load moves (D_o).

$$IMA = \frac{D_i}{D_o} \text{ or } D_o = \frac{D_i}{IMA} \qquad \text{(after rearranging equation)}$$

$$D_o = \frac{D_i}{IMA} \qquad \text{where: } D_i = 4 \text{ in.}$$
$$IMA = 4$$

Substitute known values to get:

$$D_o = \frac{4 \text{ in.}}{4} = 1 \text{ in.} \qquad \text{(Load moves up one inch, while input piston moves down 4 inches.)}$$

d. Use Formula 5 or 6 to find volume of fluid transferred. Since $\Delta V_i = \Delta V_o$, you can use either $\Delta V_i = A_i \times D_i$ or $\Delta V_o = A_o \times D_o$.

Use $\Delta V_i = A_i \times D_i$, where $A_i = 1 \text{ in}^2$, and $D_i = 4 \text{ in.}$

$$\Delta V_i = 1 \text{ in}^2 \times 4 \text{ in.}$$
$$\Delta V_i = 4 \text{ in}^3$$

Four cubic inches of liquid are displaced at the input end. You can check this by finding ΔV_o.

$$\Delta V_o = A_o \times D_o \qquad \text{where: } A_o = 4 \text{ in}^2 \text{ and } D_i = 1 \text{ in.}$$
$$\Delta V_o = 4 \text{ in}^2 \times 1 \text{ in.}$$
$$\Delta V_o = 4 \text{ in}^3 \qquad \text{(same as } \Delta V_i)$$

The hydraulic jack has many uses in industry. The hydraulic cylinder is a form of a hydraulic jack or linear mechanical force transformer that's also used widely. Similar pneumatic systems that use gas as a fluid—usually air—also are common.

WHAT'S A HYDRAULIC PRESSURE INTENSIFIER?

A *pressure intensifier* often is called a "pressure booster." It's a "true" fluid "force" transformer, since it transforms **pressure.** In uses where a low pressure must be amplified (or boosted) to a higher pressure, many of the ideas you've learned about the hydraulic jack can help you understand how pressure boosters work.

For a hydraulic jack or lift, pressures p_i and p_o were unchanged—as were the fluid volumes moved (ΔV_i and ΔV_o).

But the forces F_i and F_o, the piston face areas A_i and A_o, and the piston movements D_i and D_o were all different when compared at the input and output ends.

In the pressure booster, as you'll see, the forces F_i and F_o are the same. However, the volumes displaced (ΔV_i and ΔV_o) and the pressures (p_i and p_o) change.

Figure 7-22 shows how a pressure intensifier is made.

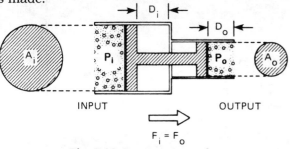

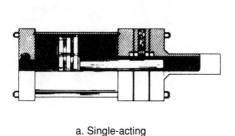

Fig. 7-22 Pressure intensifier.

On the input side, a pressure p_i acts on the piston face A_i to give a total force F_i, where $F_i = p_i \times A_i$. This force pushes the piston a distance (D_i) at the input end.

This force pushes the **same distance** ($D_o = D_i$) at the output end. The force F_i is transmitted unchanged to the output end, so that $F_o = F_i$.

At the output side, the pressure p_o equals F_o/A_o. (Recall from Unit 1 that pressure is defined as "force divided by area.")

At the input side, the pressure p_i equals F_i/A_i. Since F_o and F_i are equal, and A_o is much smaller than A_i, p_o must be greater than p_i.

Let's think through this with some values.

Suppose $F_o = F_i = 1000$ N. Let $A_i = 25$ cm^2 and $A_o = 5$ cm^2. Then:

$$p_i = \frac{F_i}{A_i} = \frac{1000 \text{ N}}{25 \text{ cm}^2} = 40 \text{ N/cm}^2$$

On the other hand,

$$p_o = \frac{F_o}{A_o} = \frac{1000 \text{ N}}{5 \text{ cm}^2} = 200 \text{ N/cm}^2$$

The pressure has been boosted by a factor of 5, from 40 N/cm^2 to 200 N/cm^2.

Figure 7-23a shows a cutaway view of a single-action hydraulic pressure booster.

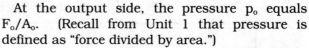

a. Single-acting   b. Double-acting

Fig. 7-23 Pressure intensifiers.

Figure 7-23b shows a double-acting pressure booster. Both are true pressure transformers. Sometimes, the low-pressure medium on the inlet end is air, while the medium at the outlet end is oil, a so-called air-over-hydraulic system. It's used when a clamping device is needed to apply high pressure on a part while it's being machined or polished.

Air-over-hydraulic intensifiers get input pneumatic pressure from a low-pressure, air-compressor line. In turn, the intensifier uses the air pressure to boost the sealed hydraulic pressure to an extremely high level. The mechanical advantage is often as high as 50 to 1.

Table 7-5, which appears on the following page, sums up the formulas that can be used to solve fluid transformer problems. You'll use several of these formulas to work a problem that appears in Example 7-M.

TABLE 7-5: PRESSURE TRANSFORMER FORMULAS FOR A PRESSURE BOOSTER

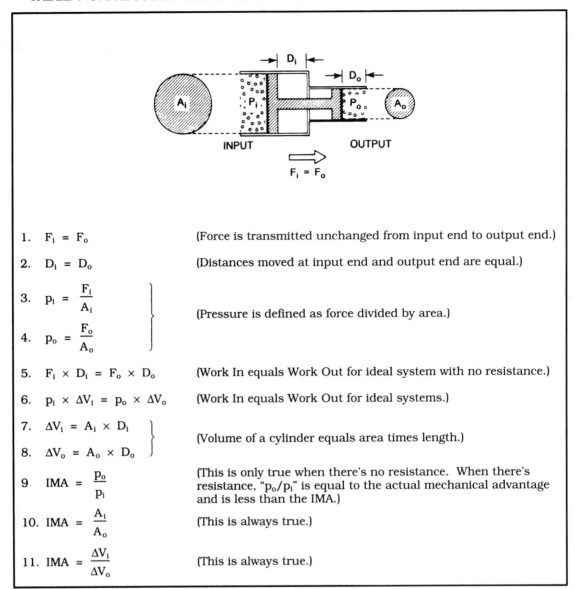

1. $F_i = F_o$ (Force is transmitted unchanged from input end to output end.)

2. $D_i = D_o$ (Distances moved at input end and output end are equal.)

3. $p_i = \dfrac{F_i}{A_i}$

4. $p_o = \dfrac{F_o}{A_o}$ (Pressure is defined as force divided by area.)

5. $F_i \times D_i = F_o \times D_o$ (Work In equals Work Out for ideal system with no resistance.)

6. $p_i \times \Delta V_i = p_o \times \Delta V_o$ (Work In equals Work Out for ideal systems.)

7. $\Delta V_i = A_i \times D_i$

8. $\Delta V_o = A_o \times D_o$ (Volume of a cylinder equals area times length.)

9. $IMA = \dfrac{p_o}{p_i}$ (This is only true when there's no resistance. When there's resistance, "p_o/p_i" is equal to the actual mechanical advantage and is less than the IMA.)

10. $IMA = \dfrac{A_i}{A_o}$ (This is always true.)

11. $IMA = \dfrac{\Delta V_i}{\Delta V_o}$ (This is always true.)

PRESSURE INTENSIFIERS ARE USEFUL IN INDUSTRY.

Pressure boosters or intensifiers are common in industry. Some of the conditions that make them ideal devices in fluid power systems are:

- No substantial heat problem is present when holding pressure over a period of time—as with pumps operating against a pressure-relief valve.
- The input can be a low-pressure fluid—gas or liquid. The amplifier pressure at the output eliminates the need for a pump to boost the pressure.
- The pressure intensifier eliminates the need for electrical equipment. So pressure boosters are safe in damp areas where electrical sparks would be dangerous.
- Pressure intensifiers are compact. This lets one pump deliver several pressures.
- Pressure intensifiers eliminate the need for expensive high-pressure controls in the primary supply system when attached directly to the tool that's being used.
- There is less spillage and less danger if a small intensifier ruptures on the high-pressure side.

Example 7-M: *Single-action Hydraulic Pressure Intensifier* ────────────

Given: An oil-type, single-action pressure booster similar to the one shown in Figure 7-23a. The pressure booster has an area ratio of $A_i/A_o = 7$. The inlet pressure $p_i = 150$ psi, the inlet stroke is 4 inches, and $A_i = 7$ in^2.

Find: a. Ideal mechanical advantage.
b. Discharge pressure p_o.
c. Volume of oil moved at inlet in one stroke.
d. Volume of oil moved at outlet in one stroke.
e. Actual discharge pressure if the intensifier is only 93% efficient.

Solution: a. Use Formula 10, Table 7-5, to find IMA.

$$IMA = \frac{A_i}{A_o} = 7$$

The ideal mechanical advantage is 7.

b. Use Formula 9, Table 7-5, to find discharge pressure p_o.

$$IMA = \frac{p_o}{p_i}, \text{ or } p_o = p_i \times IMA \qquad \text{(when equation is rearranged)}$$

$$p_o = 150 \text{ psi} \times 7 = 1050 \text{ psi}$$

c. Use Formula 7, Table 7-5, to find volume of oil moved at inlet, where $A_i = 7$ in^2 and $D_o = 4$ in.

$$\Delta V_i = A_i \times D_i$$
$$\Delta V_i = 7 \text{ in}^2 \times 4 \text{ in.} = 28 \text{ (in}^2 \times \text{in.)}$$
$$\Delta V_i = 28 \text{ in}^3$$

d. Use Formula 8, Table 7-5, to find volume of oil moved at outlet.

$$\Delta V_o = A_o \times D_o$$

First, find A_o from $A_i/A_o = 7$ and $A_i = 7$ in^2.

$$\frac{A_i}{A_o} = 7$$

$$A_o = \frac{A_i}{7} \qquad \text{(after rearranging equation)}$$

$$A_o = \frac{7 \text{ in}^2}{7} = 1 \text{ in}^2$$

Now substitute $\Delta V_o = A_o \times D_o$, where $A_o = 1$ in^2 and $D_o = 4$ in.

$$\Delta V_o = 1 \text{ in}^2 \times 4 \text{ in.}$$
$$\Delta V_o = 4 \text{ (in}^2 \times \text{in.)}$$
$$\Delta V_o = 4 \text{ in}^3 \qquad \text{(Formula 11 can also be used, with the same result.)}$$

e. Use the general equation that Eff $=$ AMA/IMA to find AMA. Then use AMA $= p_o(\text{actual})/p_i$ to find the true output pressure, where IMA = 7 and Eff = 0.93

$$Eff = \frac{AMA}{IMA}, \text{ or } AMA = IMA \times Eff$$

By substituting the values, you get:

$$AMA = 7 \times 0.93$$
$$AMA = 6.51$$

Now

$$AMA = \frac{p_o \text{ (actual)}}{p_i}$$

or

$$p_o \text{ (actual)} = AMA \times p_i \qquad \text{where: } AMA = 6.51$$
$$p_i = 150 \text{ psi}$$

$$p_o \text{ (actual)} = 6.51 \times 150 \text{ psi}$$
$$p_o \text{ (actual)} = 976.5 \text{ psi}$$

This is less than the *ideal* output pressure $p_o = 1050$ psi found in Part b.

The following exercises review the main ideas and definitions presented in this subunit, "Force Transformers in Fluid Systems." Complete each question.

1. In a hydraulic jack, **work input** equals **work output** is expressed as:
    a. $p_i \times D_i = p_o \times D_o$
    b. $p_i \times \Delta V_i = p_o \times \Delta V_o$
    c. $p_i \times (Q_v)_i = p_o \times (Q_v)_o$

2. In a hydraulic jack, pressure ____ (varies, is constant) throughout the system.

3. In a hydraulic jack, ____ (a different, the same) volume of fluid is displaced at the input end and the output end to achieve a mechanical advantage.

4. The ideal mechanical advantage (IMA) in a hydraulic jack is given by:
    a. $IMA = p_o/p_i$
    b. $IMA = \Delta V_o/\Delta V_i$
    c. $IMA = D_o/D_i$
    d. $IMA = D_i/D_o$

5. What's the ideal mechanical advantage of a hydraulic jack with an input force of 25 lb that delivers a 250-lb output force to the load? Assume no friction.

6. In Problem 5, how far must the input force act if the output force moves the load 1 inch?

7. In Problem 5, what's the ratio of the areas of the piston faces on the input and output pistons? Which piston is larger?

8. Air ____ (can, cannot) be used in conjunction with hydraulic fluid in a pressure intensifier.

9. In a pressure intensifier, while area of piston face decreases from input piston to output piston, output pressure ____ (increases, decreases) from the input to the output side of the intensifier.

10. Is the garage car lift shown more like a hydraulic jack or a pressure intensifier? Explain your answer in one or two sentences.

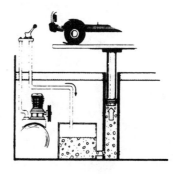

11. In a hydraulic jack, mark those quantities that **increase** from input to output, remain the **same,** or **decrease.** Use the letter, I, for increase; S for same; D for decrease.
    a. Pressure   ____
    b. Force   ____
    c. Piston movement ____
    d. Piston area   ____
    e. Volume displaced ____

12. In a pressure intensifier, mark those quantities that **increase** from input to output, remain the **same,** or **decrease.** Use the letter, I, for increase; S for same; D for decrease.
   a. Pressure ____
   b. Force ____
   c. Piston movement ____
   d. Piston area ____
   e. Volume displaced ____

13. If a pressure booster has an input piston surface of 10 in^2 and an output piston surface of 2 in^2, what would be:
   a. the ideal mechanical advantage?
   b. the discharge pressure if input pressure is 100 psi?
   c. the ratio of D_o/D_i?; of F_o/F_i?

Math Skills Laboratory

MATH ACTIVITY
Solving "Force" Transformer Problems for Fluid Systems

MATH SKILLS LAB OBJECTIVES
When you complete these activities, you should be able to do the following:
1. **Solve and interpret "force" transformer problems for fluid systems.**
2. **Relate ideal mechanical advantage to ratio of piston face areas (or diameters) on input and output sides of the transformer.**
3. **Relate actual mechanical advantage to ratio of forces (or pressures) on input and output sides of the transformer.**

LEARNING PATH
1. **Read the Math Skills Lab. Give particular attention to the Math Skills Lab Objectives.**
2. **Work the problems.**

ACTIVITY
Solving "Force" Transformer Problems for Fluid Systems

In this lab, you'll work problems that involve "force" transformers in fluid systems. The problems involve fluid pressure intensifiers, fluid jacks and cylinders.

To solve these problems, refer to the formulas in Table 7-4, "Force Transformer Formulas for a Hydraulic Jack," and Table 7-5, "Force Transformer Formulas for a Pressure Booster."

Problem 1: Given: Shop Press, Inc., makes presses like the one in this drawing. This press uses a hydraulic jack to apply force to parts that are being put together and taken apart. In fluid systems, pressure is constant throughout the volume containing the fluid. This means that "pressure input" equals "pressure output" in a hydraulic jack. The ideal mechanical advantage of a hydraulic jack can be stated as:

$$IMA = \frac{F_o}{F_i} = \frac{A_o}{A_i}$$

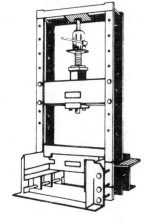

Shop press

(Suppose the input pump piston of the hydraulic jack shown in the drawing has a surface area of 1.77 in^2.)

Find: a.   Surface area of the output piston face when a 50-pound input force causes the output piston of the jack to raise a 600-pound load.

           b.   The diameter of the output piston when a 50-pound input force moves a 600-pound load.

           c.   The distance the load moves if the input piston moves 12 inches, under the conditions of b.

Solution:

Problem 2: Given: Century Construction Company has several hydraulically operated earth-moving machines. Ron Brown works as a hydraulics technician for the company. He's been asked to modify the hydraulic system of one of the machines. He'll replace a small hydraulic cylinder with a larger one. The new cylinder has a usable output piston face area of 10 in^2. The pump on this machine can provide 1500 psi of pressure in the fluid.

Find: a.   Maximum output force that can be exerted by the cylinder.

           b.   Output work done by the cylinder to move the cylinder rod out 1 ft against a load.

           c.   Input force supplied by the pump if the system's ideal mechanical advantage is 12.

           d.   Input work done by the pump (assuming there are no losses due to friction/resistance).

Solution:

Problem 3: Given: Carl Renchler is a machine repair technician for Harris Manufacturing. This company produces copper fittings. These fittings are used by plumbers and makers of heating and cooling equipment. Fitting ends are expanded so that tubes can be eased into the fittings and soldered into place. Fitting ends are expanded in a forming machine that's equipped with an air-to-water pressure booster. The pressure booster is usually operated at 125 psi air pressure. This produces a 20:1 ideal mechanical advantage.

Find: a.   The water pressure applied to the inside surface of the copper fitting.

           b.   The face area of the output piston if the input piston is 3 inches in diameter.

Solution:

Problem 4: Given: Kali-Alto Laboratories employs Kay Lewis as a standards lab technician. Part of Kay's job is to calibrate high-pressure hydraulic gages. An air-to-hydraulic pressure intensifier is used to pressurize the gages being tested and calibrated. The intensifier has a 3.5-inch-diameter input cylinder and a 0.5-inch-diameter output cylinder.

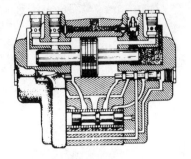

Intensifier

Find:
a. Maximum pressure available to test the gages if the air pressure is 100 psi.

b. Mechanical advantage of the pressure intensifier.

c. Overall mechanical advantage if the output pressure of the intensifier were connected to the input side of an identical pressure intensifier.

Solution:

Problem 5: Given: Power Systems makes a pressure booster that's used on the power brake system of some cars. One side of the pressure booster is linked to the intake manifold of a car's engine. When the engine runs, a vacuum forms on this side of the pressure booster. When the brake pedal is pushed down, atmospheric air pressure acts on the surface of the large piston. This increases the pressure available to push on the piston in the master cylinder of the hydraulic brake system. A research technician for this company might be asked to provide an answer to the following question.

Find: If the atmospheric pressure is 14.7 psi and the mechanical advantage of the booster is 12, what's the pressure applied to the piston of the master cylinder?

Solution:

Problem 6: Given: Williamsport Tower Company constructs and installs girder towers that support high-voltage electric power lines. These towers are very tall. They must be assembled "on site." Individual pieces of the tower are riveted together with a portable air-to-hydraulic rivet gun. The rivet gun is a pressure intensifier that uses a 50-psi air source.

Find:
a. Mechanical advantage of the pressure intensifier when a 1-in^2 output piston pushes on a rivet with 600 lb of force.

b. Area of the piston on the air input side of the pressure intensifier.

Solution:

Problem 7: Given: The sketch with this problem shows a type of air-over-hydraulic lift used by automotive garages. In this example, the air compressor is connected to a hydraulic tank that has a cross-sectional area of 7 ft². The tank is connected to a hydraulic lifting cylinder. The cylinder piston has 1 ft² of face area in contact with the oil. (Note: Air pressure above fluid in hydraulic tank serves the same function as the input piston in a hydraulic jack.)

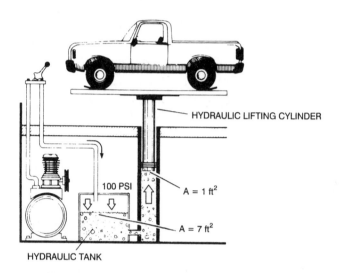

Find: a. The ideal mechanical advantage of this device.

b. The force applied to a load by the movable piston if 100 psi of air pressure is applied to oil in the tank. Remember: 1 ft² = 144 in².

c. How far up the top of the lifting piston a load moves if the oil level in the large tank moves down one foot.

Solution:

Force Transformers: The Hydraulic Jack

LABORATORY OBJECTIVES

When you've finished this lab, you should be able to do the following:

1. *Recognize that a hydraulic jack combines a simple lever and a hydraulic force transformer.*

2. *Assemble an apparatus that tests the principles of a hydraulic jack.*

3. *Find the mechanical advantage of a hydraulic jack.*

4. *Measure the input work and output work of a hydraulic jack. Find the jack's efficiency.*

LEARNING PATH

1. *Preview the lab. This will give you an idea of what's ahead.*

2. *Read the lab. Give particular attention to the Lab Objectives.*

3. *Do lab, "Force Transformers: The Hydraulic Jack."*

MAIN IDEAS

- *A force transformer changes a small force applied over a certain distance to a larger force applied over a smaller distance.*

- *The hydraulic jack is a useful force transformer that amplifies (or increases) force.*

- *The ideal mechanical advantage of a hydraulic jack equals the ratio of input displacement to output displacement.*

- *The actual mechanical advantage of a hydraulic jack equals the ratio of output force to input force.*

A hydraulic jack is a powerful force transformer. It's made of a simple, second-class lever and a hydraulic force transformer.

A hydraulic jack is used to lift heavy objects such as cars, trucks, buses, earth-moving equipment and aircraft. When a house is moved from one location to another, hydraulic jacks are needed to lift the structure from its foundation.

Even in the early 1900s, hydraulic jacks provided the force needed to lift and move a massive structure. For instance, Western Electric Company in Indianapolis, Indiana, wanted to move its 10-story home office building to another site. This office building was located in the heart of the city.

Movers used hydraulic jacks to lift the building from its foundation. Then, more than 1000 specially designed wheels were attached to the bottom of the building. Hydraulic jacks were then used to help move the building onto a street, down the street, around a corner, down another street and onto its new foundation.

Hydraulic jacks are made in many shapes and sizes. Each jack has a safety limit—or the amount of force it can safely apply. These jacks are usually made to be used in either a vertical or a horizontal position. In this lab, you'll explore several technical details about hydraulic jacks.

LABORATORY

EQUIPMENT

Hydraulic jack assembly, ½- to 1 ½-ton capacity, with pressure gage (1000-lb/in² capacity)
Spring scale, 5-lb (20-N) or 25-lb (100-N) capacity
Vernier caliper
Ruler (12 in., with centimeter scale)

PROCEDURES

Part 1: Familiarization

1. Examine the hydraulic jack. Read all instructions printed on the label fastened to the jack. Then answer these questions:
    a. What type of working fluid is used in the hydraulic jack?
    b. Are there applications in which the jack can't be used? If so, what are they?
    c. What is the rated load-limit of the jack?
    d. How many different mechanical force transformers make up the jack? Identify them.

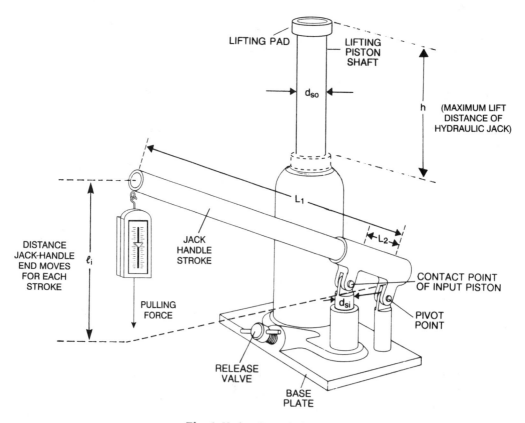

Fig. 1 Hydraulic jack dimensions.

2. Make some basic measurements. Use a ruler and a vernier caliper to measure each quantity listed in Data Table 1. Figure 1 identifies these parts. If you are unable to measure any one of these, see your teacher.

DATA TABLE 1

Diameter of: Input piston shaft $\ldots\ldots\ldots\ldots\ldots\ldots\ldots\ldots\ldots\ldots\ldots$ d_{si} =	_____
Output piston shaft $\ldots\ldots\ldots\ldots\ldots\ldots\ldots\ldots\ldots\ldots$ d_{so} =	_____
Distance end of jack handle moves for one complete stroke $\ldots\ldots\ldots\ldots$ ℓ_i =	_____
Length of measure:	
From end of jack handle to pivot point $\ldots\ldots\ldots\ldots\ldots$ L_1 =	_____
From pivot point to contact point of input piston $\ldots\ldots\ldots$ L_2 =	_____
Height (maximum) of lifting piston shaft $\ldots\ldots\ldots\ldots\ldots\ldots\ldots\ldots$ h =	_____

3. Open the release valve. Push the lifting piston to the bottom of its travel.

Part 2: *Operation and Testing*

1. Assemble the pressure stage, if necessary. Install the hydraulic jack, as shown in Figure 2a. Once you've finished the assembly, your lab setup should look like Figures 2b and 2c. Refer to Figure 2 as you complete the following steps.

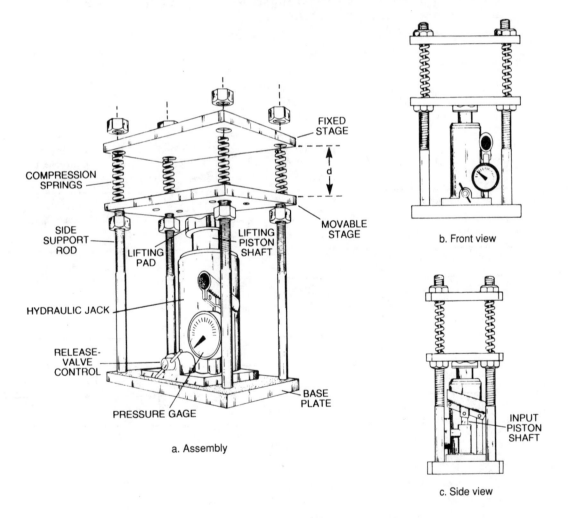

Fig. 2 Pressure stage and hydraulic jack.

2. Adjust the movable and fixed stages to be parallel. Make sure the four springs between the stages are compressed no more than $\frac{1}{16}$ inch.

3. Situate the jack so that its lifting pad is centered with respect to the movable stage.

4. Close the release valve. Insert the jack handle. Raise the handle to full elevation. Push down on the handle until the lifting pad **touches** the bottom of the movable stage and the pressure gage begins to register. Remove the jack handle.

5. Measure the distance (d_o) between the **bottom surface** of the fixed stage and the **top surface** of the movable stage. See Figure 2a. Record (to the nearest $\frac{1}{16}$ inch or nearest millimeter) under "Distance Between Stages" in Data Table 2.

6. Read the pressure gage. Record as p_o under "Pressure Gage Reading" of Data Table 2.

7. Reinsert the jack handle and move it to the top of its stroke. Put a spring balance on the end farthest from the pivot of the lever.

8. Pull down on the spring balance so that the handle moves down at a slow and steady rate. Try to keep the spring balance pulling in a direction perpendicular to the jack handle. Note the reading on the spring balance near midstroke.

9. Operate the jack handle for 4 more complete strokes, so that a total of 5 strokes are completed since the original d_o and p_o readings. Record the "average" spring balance reading near midstroke as F_1, then measure the stage separation d_1 (after 5 strokes) and pressure gage reading p_1 (after five strokes). Record in Data Table 2 in row labeled "After 5 strokes."

10. Repeat this sequence of five additional strokes, recording the measured values of d, p and F in Data Table 2. Continue until the Data Table has been filled (a total of 8 sets of readings)—or until the pressure gage reaches 700 lb/in^2.

11. Remove the jack handle. Open the release valve.

DATA TABLE 2

	Readings	Distance Between Stages	Pressure Gage Reading	Input Force from Spring Balance
1	Initial	d_o =	p_o =	
2	After 5 strokes	d_1 =	p_1 =	F_1 =
3	After 10 strokes	d_2 =	p_2 =	F_2 =
4	After 15 strokes	d_3 =	p_3 =	F_3 =
5	After 20 strokes	d_4 =	p_4 =	F_4 =
6	After 25 strokes	d_5 =	p_5 =	F_5 =
7	After 30 strokes	d_6 =	p_6 =	F_6 =
8	After 35 strokes	d_7 =	p_7 =	F_7 =

Calculations

For the following calculations, refer to Data Table 3. It's important to remember that five complete strokes were made from row to row in Data Table 2.

1. Calculate the distance of travel for the *output piston* for five stroke intervals. This distance equals the change in "Distance Between Stages" of successive readings. Record your answer in the space, "Lifting Piston Travel," in Data Table 3.

2. Calculate and record the "Average Pressure" (p_{av}) for each set of five complete strokes. Pressure average is found by taking half the sum of the pressures before and after each set of five readings—as indicated in Data Table 3.

3. Calculate the total distance moved by the *input force* for every 5 strokes of work. This distance is $5 \times \ell_i$, where ℓ_i is the value recorded in Data Table 1. This value is the same throughout, so record it in all rows of Data Table 3 under "Input Force/Travel."

DATA TABLE 3

	Lifting Piston Travel	Average Pressure P_{av}	Input Force Travel $(5 \times \ell_i)$	Average Input Force
1				
2	$d_0 - d_1 =$	$\dfrac{p_0 + p_1}{2} =$		$\dfrac{F_0 + F_1}{2} =$
3	$d_1 - d_2 =$	$\dfrac{p_1 + p_2}{2} =$		$\dfrac{F_1 + F_2}{2} =$
4	$d_2 - d_3 =$	$\dfrac{p_2 + p_3}{2} =$		$\dfrac{F_2 + F_3}{2} =$
5	$d_3 - d_4 =$	$\dfrac{p_3 + p_4}{2} =$		$\dfrac{F_3 + F_4}{2} =$
6	$d_4 - d_5 =$	$\dfrac{p_4 + p_5}{2} =$		$\dfrac{F_4 + F_5}{2} =$
7	$d_5 - d_6 =$	$\dfrac{p_5 + p_6}{2} =$		$\dfrac{F_5 + F_6}{2} =$
8	$d_6 - d_7 =$	$\dfrac{p_6 + p_7}{2} =$		$\dfrac{F_6 + F_7}{2} =$

4. Determine the "Average Input Force," taking half the sum of the Input Forces before and after each set of five strokes. Record in the corresponding column of Data Table 3.

5. Calculate the input work. Multiply the "Input Force Travel" by the "Average Input Force,"—that is, the last two columns of Data Table 3. Record the "Work In" (W_i) for readings 2 through 8 (or to the end of data taken, whichever comes first) in Data Table 4.

6. Calculate and record in Data Table 4 the volume of oil moved in the lifting piston for each reading in Data Table 3, as follows:

$$\Delta V = A \times \Delta d$$

where: ΔV = volume of oil moved in cylinder

$A = \pi r^2$ (area of circular surface of lifting piston—ask your teacher for this data)

Δd = height of oil column moved, that is, the lifting piston travel from Data Table 3

7. Copy the "Average Pressure" data from Data Table 3. List it in the correct column, reading by reading, in Data Table 4.

8. Find the output work (W_o). Multiply the "Average Pressure" (p_{av}) by the fluid volume moved (ΔV) for each listing in Data Table 4. Record the answer in the column labeled "Work Out." This is the work done to compress the springs.

9. From Data Table 4 and the equation,

$$\text{Eff} = \frac{\text{Work Out}}{\text{Work In}} \times 100\%$$

find the percentage "Efficiency" for each reading and enter in Data Table 4.

DATA TABLE 4

Readings	Work In (W$_i$)	Volume Moved (ΔV)	Average Pressure (p$_{av}$)	Work Out (W$_o$)	Efficiency (%)
2					
3					
4					
5					
6					
7					
8					

WRAP-UP

1. The hydraulic jack is made up of a simple lever and a hydraulic force transformer.
   a. What's the ideal mechanical advantage of the lever? (See Data Table 1.)
   b. Is the lever a first-class, second-class or third-class lever? Why?
   c. What's the ideal mechanical advantage of the hydraulic force transformer?
   d. What's the **overall** ideal mechanical advantage of the hydraulic jack (lever + hydraulic transformer)?

2. Are the efficiencies found for the hydraulic jack reasonable? How could you increase the efficiency?

Student Challenge

1. Why were you instructed in Step 8 of Part 2 of the *Procedures*, "to pull down on the spring balance so that the handle moves at a slow and steady rate?"

2. If you worked for a company that built the apparatus shown in Figure 2, would you need to know about spring constants? Why or why not?

Force Transformers: The Pressure Intensifier

Lab 7F2

LAB OBJECTIVES

When you've finished this lab, you should be able to do the following:

1. ***Connect two air cylinders of different diameters in such a way as to build a pressure intensifier.***

2. ***Given the diameters of the two air cylinder pistons, find the ideal mechanical advantage of the lab pressure intensifier.***

3. ***From the measurements of input/output pressure and distance traveled, find both the efficiency and the actual mechanical advantage of the lab apparatus.***

LEARNING PATH

1. ***Preview the lab. This will give you an idea of what's ahead.***

2. ***Read the lab. Give particular attention to the Lab Objectives.***

3. ***Do lab, "Force Transformers: the Pressure Intensifier."***

MAIN IDEAS

- ***A pressure intensifier is a force transformer that changes a low input pressure to a higher output pressure.***

- ***A pressure intensifier increases the mechanical advantage of a fluid system.***

Power brakes involve a fluid system that's standard on most of today's cars and trucks. Riveting machines, compactors and presses also can be powered by a fluid system. The fluid systems on these machines often use a fluid force transformer—commonly called a "pressure booster" or "pressure intensifier."

In most fluid systems, a pump is the chief source of fluid power. The pressure booster's job is similar to the job done by a fluid pump. Both develop pressure and move fluid. One difference between a pump and a pressure booster is that the basic operation of a pump results from a rotary motion. The basic operation of a pressure booster results from linear motion. (See Figure 1.)

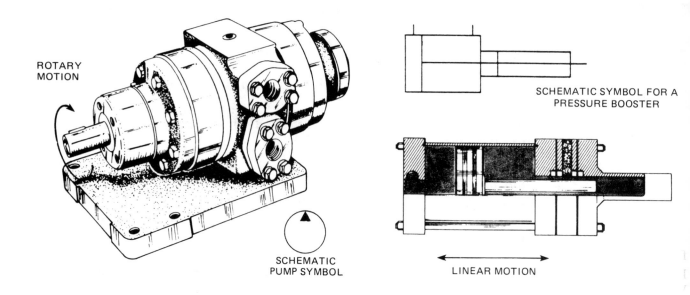

ROTARY
MOTION

SCHEMATIC
PUMP SYMBOL

SCHEMATIC SYMBOL FOR A
PRESSURE BOOSTER

LINEAR MOTION

Fig. 1 Pump and pressure booster.

Pressure boosters are like fluid cylinders. A pressure booster is just one cylinder that's operated by another cylinder. Figure 2 shows two cylinders connected to form a pressure booster. We'll label the cylinders "I" and "O" (for input and output).

In Unit 1, *Force*, you learned about the equation, $p = \frac{F}{A}$. This equation can be rearranged as $F = p \times A$. Since the cylinders are connected, the output force from the piston in cylinder "I" becomes the input force that acts on the piston in cylinder "O." Therefore, cylinder "O" becomes a pump. This pump develops higher pressure than the pump that operated cylinder "I."

Pressure boosters have many uses. For instance, pressure boosters help shape some types of metal and plastic tubing. Figure 3 shows this kind of tube-forming device.

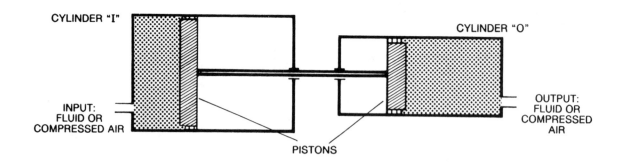

CYLINDER "I"

CYLINDER "O"

INPUT:
FLUID OR
COMPRESSED AIR

OUTPUT:
FLUID OR
COMPRESSED
AIR

PISTONS

Fig. 2 Pressure booster.

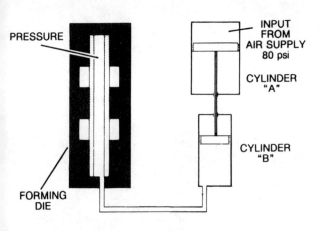

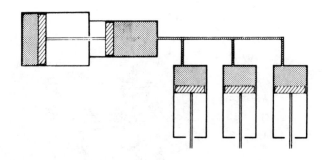

PRESSURE

FORMING DIE

INPUT FROM AIR SUPPLY 80 psi

CYLINDER "A"

CYLINDER "B"

Fig. 3 Tube-forming device.

enough, the tube in the forming-die expands and conforms to the shape of the die.

Pressure boosters also can be used to operate another cylinder or a group of cylinders, as shown in Figure 4.

Fig. 4 Pressure booster used to operate several cylinders.

In Figure 3, the tube is clamped inside a forming die. Then the tube and the high-pressure end of the pressure booster are filled with a liquid. This liquid is usually water.

Compressed air is put on the piston of cylinder A. This means force also is applied to the piston in cylinder B. Cylinder B, in turn, puts force on the water. When the force becomes large

In this lab, you'll build and test a pressure booster. You'll do this by connecting two "air" cylinders. You'll then compare the input force of one cylinder to the output force of the other cylinder.

LABORATORY

EQUIPMENT

Compressed air supply (for example, air tank with pressure gage and shutoff valve, about 6-gallon capacity, charged to 100 psi)
Pressure regulator, 0 to 30 psi
Pressure gage, two, compound type, 15 mm Hg vacuum to 30 psi pressure, with connecting tee
Air cylinders with coupling slug:
First cylinder: 1 $\frac{1}{8}$-in. diameter × 6- to 8-in. stroke
Second cylinder: $\frac{3}{4}$-in. diameter × 6-in. stroke
Support stand, rods, and clamps
Bleeder valve, two, with tees
Plastic tubing, approximately $\frac{1}{4}$" ID

PROCEDURES

Part 1: *Assembling and Aligning the Equipment*

1. Set up the lab equipment, as shown in Figure 5. Many methods can be used to fasten the cylinders to the support stand. Your teacher may have special guidelines for setting up the equipment. Read through the following steps. Refer to Figure 5 to help you understand each step.

2. Record the diameters of the driving and driven cylinder pistons in your Data Table.

3. Connect the rods of each cylinder with a coupling. The larger cylinder will be the input (driving) cylinder. The smaller cylinder will be the output (driven) cylinder.

4. Adjust the distance between the cylinders so that the cylinder rod of the driven cylinder can travel its full stroke.

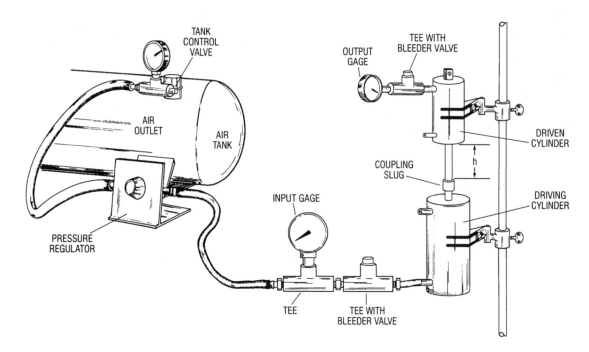

Fig. 5 Lab setup.

5. Connect and secure the cylinders. Adjust the alignment of the cylinder rods. Rods are aligned properly when they slide into and out of the cylinders easily without binding.

6. Move the driving cylinder rod back into its cylinder as far as it will go. This will cause the driven cylinder rod to position itself at the start of its stroke.

7. Connect a section of hose from the control valve of the air supply to the inlet of the pressure regulator. Use other sections of hose to connect the pressure-regulator outlet to a tee fitting that holds the input pressure gage, and then to another tee that holds a bleeder valve. Use a short length of tubing to connect the other end of the tee to the input side of the driving cylinder, as shown.

8. Connect the second pressure gage to the output of the driven cylinder, using a tee to include another bleeder valve, as shown.

Part 2: *Conducting the Lab*

1. Turn the pressure-regulator knob fully counterclockwise (ccw). This should **prevent** any airflow from the tank to the cylinders. Open the air-tank control valve. Make sure that air from the tank isn't leaking through the regulator.

2. Measure the initial distance h_i from the bottom of the driven cylinder to the top of the cylinder rod coupling. (See Figure 5.) Record this value in your Data Table.

3. Slowly turn the pressure-regulator knob clockwise (cw) to start flow from the tank to the cylinders. Stop turning the knob when the input pressure gage starts to move.
   a. Did the cylinder rods move?
   b. If not, why hasn't the input piston moved?

4. Slowly increase the pressure in the driving cylinder. To do so, turn the pressure-regulator knob farther clockwise. Note the pressure reading on the input pressure gage when the cylinder rods **first** move. Label this $(p_i)_o$ and record the value in your Data Table.

5. Continue to increase the pressure in the driving cylinder **until** the pressure reading on the output pressure gage **stops increasing**.

6. At this point, record the readings of the input gage (p_i) and output gage (p_o) in the Data Table.

7. Measure the final distance h_f between the bottom of the driven cylinder and the top of the cylinder rod coupling. Record this value in the Data Table.

Diameter of:
    Driving-cylinder piston (d_i) = _____
    Driven-cylinder piston (d_o) = _____
Initial Readings:
    Distance (h_i) = _____
    Input pressure from Part 2, Step 4 ($p_i)_o$ = _____
Final Readings:
    Input pressure from Part 2, Step 6 (p_i) = _____
    Output pressure from Part 2, Step 6 (p_o) = _____
    Distance (h_f) = _____

CALCULATIONS

1. Find the area (A) of each piston in the two cylinders.
    a. Driving-cylinder area (A_i).
        $A_i = \frac{1}{4}\pi \, d_i^2 = (0.25)(3.14)(\underline{})^2 = \underline{}$
    b. Driven-cylinder area (A_o).
        $A_o = \frac{1}{4}\pi \, d_o^2 = (0.25)(3.14)(\underline{})^2 = \underline{}$

2. Find the ideal mechanical advantage (IMA) of the lab setup.
        $IMA = \dfrac{A_i}{A_o} = (\underline{})/(\underline{}) = \underline{}$

3. Find the change in volume of each piston/cylinder using this equation:
        $\Delta V = A \Delta \ell$ where: $\Delta \ell = h_i - h_f$
                    $\Delta \ell = (\underline{}) - (\underline{}) = \underline{}$
    a. $(\Delta V)_i = A_i \Delta \ell = (\underline{})(\underline{}) = \underline{}$
    b. $(\Delta V)_o = A_o \Delta \ell = (\underline{})(\underline{}) = \underline{}$

4. Calculate % efficiency of the lab setup.
        $Eff = \dfrac{\text{Work Out}}{\text{Work In}} \times 100\%$
    a. Fluid Work In:
        $W_i = p_i(\Delta V)_i = (\underline{})(\underline{}) = \underline{}$
    b. Fluid Work Out:
        $W_o = p_o(\Delta V)_o = (\underline{})(\underline{}) = \underline{}$
        $Eff = \dfrac{W_o}{W_i} \times 100\% = (\underline{})/(\underline{}) \times 100\% = \underline{}$

1. What conclusions can you make about the pressure needed to start rod movement? (See Steps 3 and 4 of **Part 2** of the lab.)
2. Use the equation, AMA = Eff × IMA, to find the actual mechanical advantage of the pressure intensifier. (Be sure that you use "Eff" as a **decimal number** in this equation.)

Student Challenge ────────────────────────────────────

How would the efficiency and AMA be affected if the value of pressure recorded in **Part 2,** Step 4, were subtracted from the "p_i" recorded in Step 6?

Review

1. *View and discuss the video, "Force Transformers in Fluid Systems."*
2. *Review the Objectives and the Main Ideas of the print materials in this subunit.*
3. *Your teacher may give you a test over Subunit 3, "Force Transformers in Fluid Systems."*

SUBUNIT 4

Force Transformers In Electrical Systems

SUBUNIT OBJECTIVES

When you've finished reading this subunit and viewing the video, "Force Transformers in Electrical Systems," you should be able to do the following:

1. *Describe how a voltage transformer is used to "step up" or "step down" AC voltage.*

2. *Explain the relationship between input power and output power for a voltage transformer.*

3. *Explain the relationship between voltage in, voltage out, and number of wire windings in a voltage transformer.*

4. *Find the "electrical" advantage of a voltage transformer.*

5. *Find the efficiency of a voltage transformer.*

6. *Measure the "electrical" advantage of a voltage transformer.*

7. *Identify workplace applications where technicians use electrical transformers.*

LEARNING PATH

1. *Read this subunit, "Force Transformers in Electrical Systems." Give particular attention to the Subunit Objectives.*

2. *View and discuss the video, "Force Transformers in Electrical Systems."*

3. *Participate in class discussions.*

4. *Watch a demonstration about electrical force transformers.*

5. *Complete the Student Exercises.*

MAIN IDEAS

- *Electrical transformers can step up or step down AC voltage and current.*

- *Varying the ratio of wire turns on the input and output coils of the transformer changes the electrical advantage.*

- *Transformers can transfer electricity economically over long distances.*

- *Voltage transformer electrical advantage equals the specific ratios of voltages, currents or wire turns.*

- *Electrical transformer efficiency equals the ratio of power out/power in, just as for other transformers.*

WHAT'S AN ELECTRICAL 'FORCE' TRANSFORMER?

The purpose of an electrical transformer is to increase or decrease the *voltage* available from an alternating current (AC) source. Figure 7-24 shows a simple transformer. The transformer has three principal parts: a primary coil or winding connected to a source of alternating current; a secondary coil or winding connected to a load; and a soft iron core.

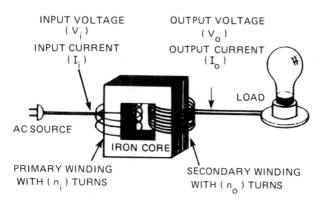

INPUT VOLTAGE (V_i)
INPUT CURRENT (I_i)
OUTPUT VOLTAGE (V_o)
OUTPUT CURRENT (I_o)
LOAD
AC SOURCE
IRON CORE
PRIMARY WINDING WITH (n_i) TURNS
SECONDARY WINDING WITH (n_o) TURNS

Fig. 7-24 A simple transformer.

When alternating current flows in the primary windings, a changing magnetic field is set up. The iron core concentrates the magnetic field in the iron. The magnetic field changing in the iron core creates voltage and current flow in the secondary windings. The magnitude of the voltage set up in the secondary winding depends on the ratio of the number of wire turns around the iron core in each of the two coils.

There are two main types of transformers. A transformer that *increases* output voltage compared with input voltage is called a *step-up transformer*. A transformer that *decreases* output voltage compared with input voltage is called a *step-down transformer*.

The complete theory behind the transformer is complicated. It involves the magnetic fields produced within the iron core by the changing current in the windings. However, the theory proves that the voltage across the primary and secondary windings is proportional to the number of turns, or windings, on each. The equation is as follows:

$$\frac{V_o}{V_i} = \frac{n_o}{n_i}$$

where: V_o = output or secondary voltage
n_o = wire turns on secondary winding
V_i = input or primary voltage
n_i = wire turns on primary winding

If you rearrange the equation and isolate V_o, you get:

$$V_o = \left(\frac{n_o}{n_i}\right) \times V_i$$

The voltage or *forcelike quantity* at the source (V_i) is multiplied by the ratio of turns (n_o/n_i) to give the voltage at the output or load. This ratio (n_o/n_i) is therefore the ideal electrical advantage of the transformer:

$$IEA = \frac{n_o}{n_i}$$

This ratio of turns depends on the physical makeup of the coils of wire. Once found, the ratio doesn't change. That's why the equation $IEA = n_o/n_i$ gives a fixed, ideal electrical advantage that doesn't depend on current or voltage ratios.

Current and voltage ratios change. They change because both current and voltage change, depending on how much resistance is present.

Because voltages change when resistance is present, the ratio of output voltage to input voltage doesn't give the *ideal* electrical advantage, as you've seen many times. Instead, the ratio V_o/V_i gives the *actual* electrical advantage.

Actual mechanical advantage is always lower than ideal mechanical advantage, as you know. In mechanical transformers, you can blame the lower values of actual mechanical advantage (AMA) on the presence of *mechanical* resistance, or friction. In electrical transformers, you can blame the lower values of electrical advantage on the presence of *electrical* resistance. So you have two equations for electrical advantage.

Ideal: $$IEA = \frac{n_o}{n_i}$$

where: n_i = number of input coil windings (primary)
n_o = number of output coil windings (secondary)

Actual: $$AEA = \frac{V_o}{V_i}$$

where: V_o = output voltage
V_i = input voltage

IEA is greater than AEA.

Note: While one can still use the words ***mechanical advantage***—in an *analogous sense*—when describing the action of electrical transformers we have chosen to use ***electrical advantage*** to avoid confusion. Thus AMA becomes AEA and IMA becomes IEA.

HOW ABOUT INPUT AND OUTPUT POWER FOR A VOLTAGE TRANSFORMER?

In previous subunits on "force" transformers, you learned that Work In = Work Out when no friction or resistance is present. For electrical transformers, it's **easier** to work with input power and output power. That's because you usually measure current (I) and voltage (ΔV) in electrical circuits.

Power (P) equals *voltage times current* ($\Delta V \times I$), as you know from Unit 6. So for electrical transformers, instead of "Work In = Work Out" use "Power In = Power Out." And remember, this ideal balance is true when there's no electrical resistance. But that hardly ever happens.

For an ideal voltage transformer:
$$\text{Power In} = \text{Power Out}$$
$$V_i \times I_i = V_o \times I_o$$

where:
V_i = input voltage (primary coil)
I_i = input current (primary coil)
V_o = output voltage (secondary coil)
I_o = output current (secondary coil)

For this equation to balance, a voltage step-up must be accompanied by a current step-down. The equation also will balance when there's a voltage step-down and a current step-up. As a result, an electric transformer can be a current transformer as well as a voltage transformer. For example, if an input consists of a low current at a high voltage, you can choose a transformer whose output consists of a high current at a low voltage.

Table 7-6 sums up the important formulas for a voltage transformer.

TABLE 7-6: USEFUL FORMULAS FOR VOLTAGE TRANSFORMERS

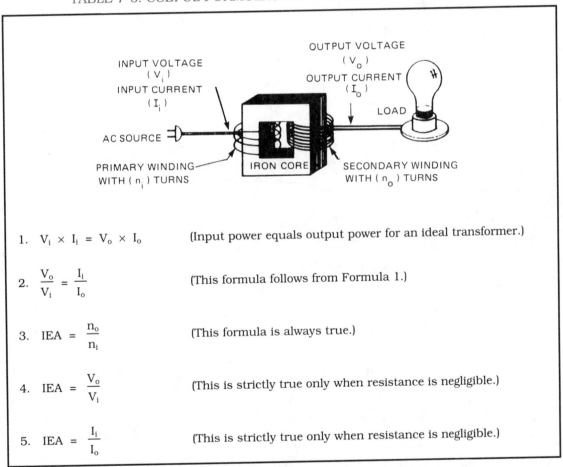

1. $V_i \times I_i = V_o \times I_o$ (Input power equals output power for an ideal transformer.)

2. $\dfrac{V_o}{V_i} = \dfrac{I_i}{I_o}$ (This formula follows from Formula 1.)

3. $\text{IEA} = \dfrac{n_o}{n_i}$ (This formula is always true.)

4. $\text{IEA} = \dfrac{V_o}{V_i}$ (This is strictly true only when resistance is negligible.)

5. $\text{IEA} = \dfrac{I_i}{I_o}$ (This is strictly true only when resistance is negligible.)

Now apply the ideas and the formulas in Table 7-6 to several practical problems. Examples 7-N and 7-O both involve voltage transformers. Example 7-N shows how a step-down voltage is achieved in a doorbell. Example 7-O shows how a step-down transformer can be used to increase current.

Example 7-N: *Voltage Step-down Transformer* —————————————

Given: An electrical doorbell transformer has 200 turns on the primary, or input, coil. It has 20 turns on the secondary, or output coil. The source voltage is 120 volts.

Find: The output voltage available to operate the doorbell.

Solution: Use Formulas 3 and 4 in Table 7-6 to get $\dfrac{n_o}{n_i} = \dfrac{V_o}{V_i}$.

Rearrange the formula, $\dfrac{n_o}{n_i} = \dfrac{V_o}{V_i}$, to isolate V_o. Then,

$$V_o = \frac{n_o}{n_i} \times V_i.$$

where: $n_o = 20$
$n_i = 200$
$V_i = 120 \text{ V}$

Substitute values in the formula.

$$V_o = \frac{20}{200} \times 120 \text{ V}$$

$$V_o = \left(\frac{20 \times 120}{200}\right)V$$

$$V_o = 12 \text{ V}$$

The voltage has been stepped down from 120 V to 12 V.

Example 7-O: *Increasing Current With a Transformer* —————————————

Given: The input current to a transformer is 1 ampere at 2000 volts.

Find: The output current at 100 volts. (Assume 100% efficiency in the transformer.)

Solution: Use Formula 1, Table 7-6.

$$V_i \times I_i = V_o \times I_o$$

Rearrange the formula to solve for I_o.

$$I_o = \frac{V_i}{V_o} \times I_i$$

where: $V_i = 2000 \text{ V}$
$V_o = 100 \text{ V}$
$I_i = 1 \text{ A}$

$$I_o = \frac{2000 \text{ V}}{100 \text{ V}} \times 1 \text{ A}$$

$$I_o = \left(\frac{2000 \times 1}{100}\right)\left(\frac{\cancel{V} \times A}{\cancel{V}}\right) \qquad \text{(Cancel units of V.)}$$

$$I_o = 20 \text{ A}$$

Output current has been stepped up from 1 A to 20 A.

HOW DO WE FIND ELECTRICAL ADVANTAGE FOR VOLTAGE TRANSFORMERS?

Table 7-6 listed three useful formulas for electrical advantage of voltage transformers. They involved ratios of coil windings, voltages or currents.

You can use these formulas in several practical situations. Example 7-P shows how to find the ideal electrical advantage in a transformer. Example 7-Q deals with efficiency of a particular transformer.

Example 7-P: *Electrical Advantage in Voltage Transformers*

Given: The voltage transformer shown.

Find: a. Ideal electrical advantage
              using coil-turn ratios.

           b. Ideal electrical advantage
              using voltage ratios.

           c. Ideal electrical advantage
              using current ratios.

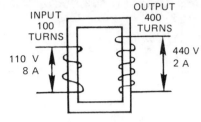

Solution: a. Use Formula 3, Table 7-6.

$$\text{IEA} = \frac{n_o}{n_i} \qquad \text{where: } n_o = 400 \text{ and } n_i = 100$$

$$\text{IEA} = \frac{400}{100} = 4$$

b. Use Formula 4, Table 7-6.

$$\text{IEA} = \frac{V_o}{V_i} \qquad \text{where: } V_o = 440 \text{ V and } V_i = 110 \text{ V}$$

$$\text{IEA} = \frac{440 \text{ V}}{110 \text{ V}} = 4 \qquad\qquad \text{(Cancel units of V.)}$$

c. Use Formula 5, Table 7-6.

$$\text{IEA} = \frac{I_i}{I_o} \qquad \text{where: } I_i = 8 \text{ A and } I_o = 2 \text{ A}$$

$$\text{IEA} = \frac{8 \text{ A}}{2 \text{ A}} = 4 \qquad\qquad \text{(Cancel units of A.)}$$

Note: For an actual transformer with resistance, the results for Parts b and c would
be somewhat less than 4.

To solve the problem in Example 7-Q, use an equation like one you learned earlier.
That equation expressed efficiency in terms of ideal mechanical advantage and actual
mechanical advantage. Recall that this equation was Eff = (AMA/IMA) × 100%.

Example 7-Q: *Voltage Transformer Efficiency*

Given: A voltage transformer is rated as having an ideal electrical advantage of IEA = 4.
           When a 110-volt source is applied to the input coil, the output voltage is 400 volts.

Find: The transformer efficiency.

Solution: Use the equation for the actual electrical advantage (AEA):

$$\text{AEA} = \frac{V_o \text{ (actual)}}{V_i} \qquad \text{where: } V_o \text{ (actual)} = 400 \text{ V and } V_i = 110 \text{ V}$$

Substitute in values.

$$\text{AEA} = \frac{400 \text{ V}}{110 \text{ V}}$$

$$\text{AEA} = 3.64$$

The actual electrical advantage is 3.64—less than the ideal electrical advantage of 4.
Percent efficiency is given by:

$$\text{Eff} = \frac{\text{AEA}}{\text{IEA}} \times 100\% \qquad \text{where: } \text{AEA} = 3.64 \text{ and IEA} = 4.0$$

$$\text{Eff} = \frac{3.64}{4.0} \times 100\%$$

$$\text{Eff} = 91\%$$

Note: In a well-designed transformer, the losses are very small. Efficiency would be
around 97% or better. That's because voltage transformers are *very* efficient.

WHAT ARE SOME USES FOR ELECTRICAL TRANSFORMERS?

There are cases where high voltages are needed. In other situations, low voltages are more useful. High voltage is used in power transmission lines. It keeps heat losses due to line resistance to a minimum. Low voltage is better in most household appliances. It makes them more safe.

In addition, transformers are used to lower the voltage and step up the current in arc welders to produce the necessary heat (I^2R losses). All these voltages are produced by the proper use of step-up and step-down transformers. A few examples are shown in the electrical power distribution system in Figure 7-25.

Examine Figure 7-25 closely. The following numbers are keyed to the figure.

1. Power is generated at 20 kV (20,000 volts). Power enters a step-up transformer before it's sent out over the transmission lines at 138 kV (138,000 V).

2. A transmission substation for a large area steps down the voltage for that area to 69 kV.

3. Industrial sites and large users have substations to step down the voltage further to meet requirements. These requirements vary.

4. Within a distribution area, the voltage is stepped down from 69 kV to a line voltage of 13.8 kV.

5. The power pole transformer outside a home steps down the voltage from 13.8 kV to 120 and/or 240 V.

6. Inside your home, you may use a transformer to step down 120 V to 12 V to operate the doorbell or some other appliance at different voltages.

7. In the industrial site, there may be a transformer that lowers the voltage and steps up the current to do welding.

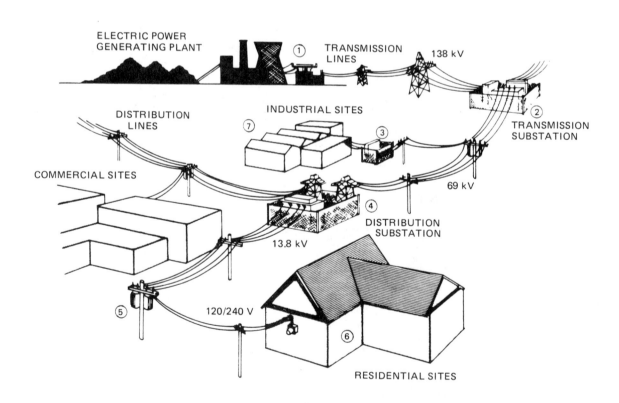

Fig. 7-25 Power from an electric generating plant.

SUBUNIT 4
Student Exercises

The following exercises review the main ideas and definitions presented in this subunit, "Force Transformers in Electrical Systems." Complete each question.

1. Electrical transformers can step up or step down ____. (Select one answer.)
   a. voltage only.
   b. current only.
   c. both voltage and current.
   d. neither voltage nor current.

2. Find the ideal electrical advantage for a transformer that has n_i = 400 turns, n_o = 900 turns, V_i = 8 V, and I_i = 0.4 A.

3. For Problem 2 above, find I_o and V_o, if resistance is considered minimal.

4. An ideal transformer has I_i = 2 A and I_o = 0.1 A. Find the wire-turns ratio (n_o/n_i) in the secondary and primary coils.

5. An ideal transformer has n_i = 1000 turns, n_o = 200 turns, I_i = 0.8 A, and V_i = 10 V. Find I_o and V_o.

6. In one or two sentences, explain why a wire-turn ratio—rather than a voltage or current ratio—is used to find ideal electrical advantage for electrical transformers.

Math Skills Laboratory

MATH ACTIVITY
Solving "Force" Transformer Problems in Electrical Systems

MATH SKILLS LAB OBJECTIVES
When you complete these activities, you should be able to do the following:
1. **Solve and interpret "force" transformer problems in electrical systems.**
2. **Distinguish between step-up and step-down transformer actions for both voltage and current.**

LEARNING PATH
1. **Read the Math Skills Lab. Give particular attention to the Math Skills Lab Objectives.**
2. **Work the problems.**

ACTIVITY
Solving "Force" Transformer Problems in Electrical Systems
In this lab, you'll solve problems that involve "force" transformers in electrical systems. You'll work problems that involve electrical transformers, where both current and voltage may be stepped up or stepped down.

To solve these problems, refer to the formulas in Table 7-6, "Useful Formulas for Voltage Transformers."

Problem 1: Given: Watson Manufacturing produces electrical components. These include capacitors, resistors, transistors, triacs, silicon-controlled rectifiers and small transformers. Rita Baxter works as a technical sales representative for the company. An important part of her job involves matching the company's products to the needs of customers. For one application, Rita suggested that her customer use a transformer that has 200 turns on the primary (input) winding and 400 turns on the secondary (output) winding.

Find: a. Output voltage of the transformer when the input coil is connected to 110 volts AC.

     b. Voltage step-up ratio (ideal electrical advantage) for the transformer.

     c. Ratio of Current Out to Current In for the transformer.

Solution:

Problem 2:   Given: One of the products made by Watson Manufacturing is a voltage transformer that has an ideal electrical advantage of 6.

   Find: a.   Number of turns in the output coil if the input coil has 50 turns.

   b.   Is this a step-up or step-down voltage transformer?

   c.   What's the output voltage if the input voltage is 60 V AC?

   Solution:

Problem 3:   Given: "Toroid" transformers (wire coil transformer) often are used with three-phase power lines to find the current flowing in each line. Some toroids are permanently installed on the individual lines of switch gears. Others—portable types—clamp around the line. The figure shows two of these "ampere meter" transformers. The ammeter is calibrated in terms of the ratio of turns of the wires. This way, a small AC current in the ammeter circuit shows the true AC current in the input conductor.

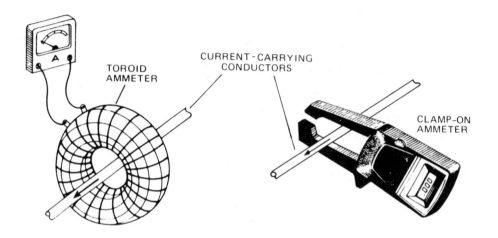

TOROID
AMMETER

CURRENT-CARRYING
CONDUCTORS

CLAMP–ON
AMMETER

   Find: a: The ideal electrical advantage for the in-line toroid transformer if the toroid has 92 turns and the input, single-line conductor passing through the toroid is considered to have a turn value of n = 1.

   b.   The ammeter indicates 15.0 amperes of current in the line. What's the actual output current of the toroid flowing to the ammeter?

   c.   When the conductor carries 15 amperes, it has a line voltage of 220 volts. Consider this the voltage in the "primary" winding. What's the voltage across the toroid terminals—that is, the voltage across the "secondary" windings?

   Solution:

Problem 4: Given: A linear variable differential transformer (LVDT) is composed of a primary coil and two secondary coils, as shown in the drawing. Inside the hollow core is a sliding armature that causes a voltage across the primary windings to appear (proportionally) across the secondary windings. The secondary windings produce equal and opposite voltages when the armature is located midway between the coils (null point). At that position, there is no resultant voltage. Moving the armature either way causes an inbalance and a positive or negative voltage across the secondary coils. Thus, measuring the secondary voltage gives a way to determine the position of the sliding armature.

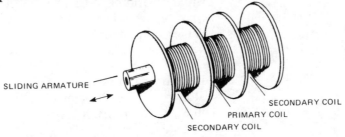

SLIDING ARMATURE

SECONDARY COIL

PRIMARY COIL

SECONDARY COIL

Find:

a. The ideal electrical advantage possible for the LVDT when the primary coil has 25 turns and each secondary coil has 300 turns.

b. The maximum output current if the input voltage is 110 V at 3 amp and the LVDT is 100% efficient.

c. The output voltage for the conditions given in a and b.

d. The output wattage.

Solution:

Problem 5: Given: Line distribution transformers that reduce power-line voltage to house-line voltage must be able to supply 120 volts to the house entrance line. That is the case regardless of the line voltage loss on the input side because of how far the transformer is from the power substation. A 2400 V-to-120 V ratio transformer may not have 2400 V available due to power-line resistance. Multiple taps are provided on the primary (input) windings. These are provided to give different numbers of turns (n_i). That's done to provide 120 volts of house-line voltage even when the distribution-line voltage is less than 2400 V. (See drawing.) Notice that the secondary voltage remains 120 volts and has 300 turns $(n_o = 300$ turns) in all four cases. The transformer is considered to be 100% efficient.

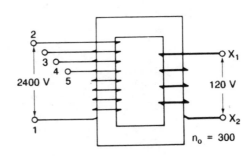

HV tap	Primary Voltage	Secondary Voltage
1-2	2400 V	120 V
1-3	2292 V	120 V
1-4	2184 V	120 V
1-5	2076 V	120 V

Find: The number of turns on the high-voltage (input or primary) side of the transformer for the taps given in Parts a-d below.

a. Tap 1-2
b. Tap 1-3
c. Tap 1-4
d. Tap 1-5

Solution:

Problem 6: Given: In an autotransformer, the secondary winding is really part of the primary winding. In effect, this means there's no need for a separate secondary winding. That makes the transformer lighter, smaller and cheaper than standard transformers of equal power output. Autotransformers are used to start induction motors, to regulate transmission-line voltages and to transform voltages when the ratio is close to 1. Ratios rarely exceed 5 to 1. An autotransformer is shown in the drawing. The primary (input) coil (n_i) contains 1000 turns.

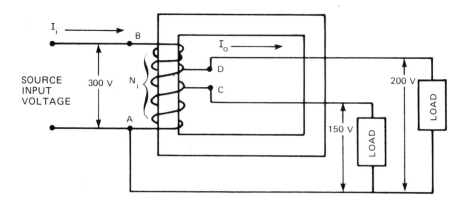

Find: a. The number of turns in the secondary (output) coil when the input voltage is 300 V and output voltage is 200 V. (Secondary taps are A and D.)

b. The number of turns in the secondary (output) coil when input voltage is 300 V and output voltage is 150 V. (Secondary taps are at A and C.)

c. The percent of the primary coil that's "tapped-off" as a secondary coil in

(1) Part a above.
(2) Part b above.

Solution:

Electrical Transformers

LABORATORY OBJECTIVES

When you've finished this lab, you should be able to do the following:

1. *Use a voltage transformer to step down input voltage.*
2. *Use a voltage transformer to step up input voltage.*
3. *Find the ideal and actual electrical advantage of a transformer.*
4. *Find power in, power out—and the operating efficiency of a simple transformer.*

LEARNING PATH

1. *Preview the lab. This will give you an idea of what's ahead.*
2. *Read the lab. Give particular attention to the Lab Objectives.*
3. *Do lab, "Electrical Transformers."*

MAIN IDEAS

- *An electrical transformer is a "force" transformer that increases or decreases input voltage.*
- *Ideal electrical advantage of an electrical transformer is determined by the ratio of wire turns on the output coil to input coil.*
- *Actual "mechanical" advantage of an electrical transformer equals the ratio of V_{out}/V_{in}. <u>Actual</u> electrical advantage is usually less than the <u>ideal</u> electrical advantage because of circuit resistance and core losses.*
- *When an electrical transformer is used to step up voltage, it's called a "step-up transformer."*
- *When an electrical transformer is used to step down voltage, it's called a "step-down transformer."*

You're being affected by electrical and electronic devices even while you read this page. What kind of light is falling on these words? Chances are, it isn't sunlight. The clothes you're wearing probably were made with the help of an electrically powered machine—right down to the shoes on your feet. The food you'll eat at your next meal probably was processed by an electrically powered machine. And your food may have been cooked electrically as well.

Even a self-sufficient hermit, living off the land, usually finds a way to get electricity. Electricity has become a basic necessity of modern life. Manufacturing depends upon electricity. Word processing and other forms of communication depend on electricity and electrical devices.

The list goes on and on. Therefore, technicians must know how to maintain, troubleshoot and repair electrical devices.

In the "guts" of most types of electrical devices, you'll find electrical transformers. Transformers are used in televisions to step up the voltage that operates the picture tube and to step down the voltage that operates the audio portion of the TV set. Transformers are also used in radios, tape players, turntables, doorbells, fluorescent lights, automotive ignition systems, lasers, and radio transmitters.

You've learned that a transformer is made of two separate coils of wire wrapped around a metal core. You've learned also that a **step-up** transformer has more coils (turns of wire) in the secondary (output) coil than in the primary or (input) coil. You know that a **step-down**

transformer has fewer coils in the secondary than in the primary.

However, there are some transformers, called "multi-tap transformers," that have more than one "secondary" coil. A multi-tap transformer can be thought of as a multiple-output transformer. This transformer is used when a single input voltage—but more than one output voltage or current—is required. A television uses a transformer to step up and step down the voltage for different parts of the TV set. Figure 1 shows a transformer that has more than one secondary coil. As before, the number of turns in each secondary coil compared to the number of turns in the primary coil determines if that "tap" is a step-up or step-down tap.

Another common type of transformer is an **isolation** transformer. This transformer has the **same** number of turns in the secondary and primary coils. Some types of circuits use AC and DC voltage at the same time. Figure 2 shows an example of a circuit that uses an isolation transformer. Transformers can change only AC voltage and current. That means an isolation transformer can be used to isolate DC voltage and current signals. The isolation transformer then allows AC signals to go from primary to secondary but blocks any DC signals in the primary.

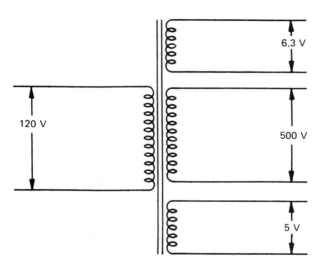

a. Schematic

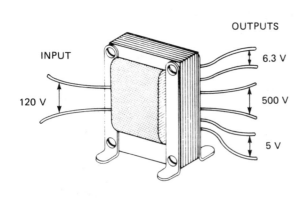

b. Pictorial

Fig. 1 Multiple-tap transformer.

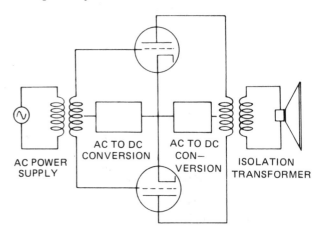

Fig. 2 Isolation transformer used to isolate DC voltage from AC voltage.

There's another good use for the isolation transformer. It can protect sensitive electronic equipment—such as computers—from sudden voltage spikes. The isolation transformer allows line-frequency signals to pass. But rapid- and short-duration spike pulses are safely reduced.

For this lab, you'll use an electrical transformer to step up and step down input voltage. You'll measure power in and power out. Then you'll find the efficiency of each transformer.

LABORATORY

EQUIPMENT

Transformer assembly with n_o/n_i ratio known (vendor supplied)
Variable AC power source, 0-25 V AC
Digital multimeter (DMM), with probes, 10 A, 0.1 mV
Solderless breadboard
Hookup wire, minimum 22 AWG
Load resistors, two, for example, 10 Ω and 100 Ω, 5 watts or higher

PROCEDURES

Part 1: *Testing a Voltage Step-down Transformer*

In the following procedures, we'll refer to the coil with the lesser number of turns as coil A; the other is coil B. (Number of turns for each coil to be provided by your teacher.)

1. First, use the transformer to "step down" the voltage. To do this, let coil B act as the input (primary) and coil A (secondary) as the output.

2. Connect the wires from coils A and B and the two resistors, R_o and R_i, to the solderless breadboard as shown in Figure 3. Study both the pictorial diagram (3a) and the schematic diagram (3b) to be sure your connections are correct.

3. Connect points "a" and "b" on the solderless breadboard to the AC power source. See Figure 3. Be sure power supply is OFF.

4. Connect a DMM between points "c" and "e." See Figure 3. Adjust the DMM to read AC volts. Select the smallest range that will read 10 volts.

5. Now turn the AC power source ON. Adjust the voltage so that the DMM across points "c" and "e" registers 10 volts. (This means that there's a 10-volt drop across the input resistor [R_i] and coil B.) Leave the AC power source at this setting.

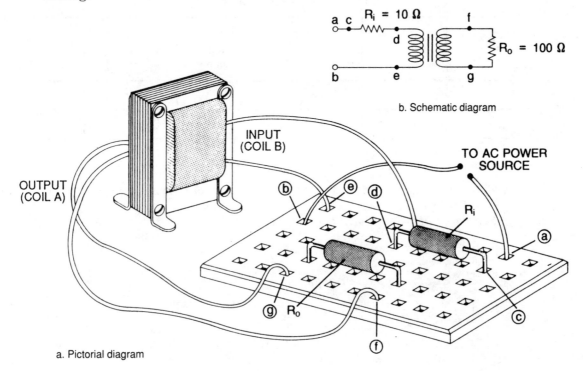

b. Schematic diagram

a. Pictorial diagram

Fig. 3 Lab setup.

6. Move the leads of the DMM across points "d" and "e" to measure the voltage across the input coil B. Read this voltage. Record in Row 1 of Data Table 1, under column "V_i."

7. Next, use the DMM to measure the voltage across coil A (output) between points "f" and "g." Read this voltage. Record in Row 2 of Data Table 1, under column "V_o."

8. Use the DMM to measure the voltage across resistor R_i, between points "c" and "d." Call this reading "V_R-in." Record in Row 1 of Data Table 1. (This measurement will allow us to find the input current.)

9. Turn the AC power source OFF.

Part 2: *Testing a Voltage Step-up Transformer*

1. To use the transformer to step up voltage, switch the connections of coils A and B. Connect the ends of coil A between points "d" and "e" on the breadboard and the ends of coil B between points "f" and "g." Now coil A is the input, or primary, coil. Coil B is the output, or secondary, coil.

2. Turn the output control of the AC power source to a "minimum" setting. Then turn source ON. Adjust the AC power source so that the DMM—connected across points "c" and "e"—now measures 5 volts. **Note:** Do not exceed 5 volts, else you may overload R_o.

3. Now connect the DMM leads across points "d" and "e." Measure the voltage drop across input coil A. Read the indicated voltage. Record in Row 2 of Data Table 1, under column "V_i."

4. Move the DMM leads to points "f" and "g" to measure the output voltage across coil B. Read the indicated voltage. Record in Row 2 of Data Table 1, column "V_o."

5. Move the DMM leads to points "c" and "d." Measure the voltage drop across resistor "R_i." Call this "V_R-in." Record in Row 2 of Data Table 1.

6. Turn the AC power source OFF.

DATA TABLE 1

Row No.	Primary Coil (input)	Secondary Coil (output)	Input Voltage V_i (volts)	Output Voltage V_o (volts)	Voltage Across R_i, V_R-in (volts)
1	B	A			
2	A	B			

Calculations

1. Calculate the turns-ratio n_o/n_i for the two types of transformers **from the data of Data Table 1**.

   Row 1 Data: Step-down transformer.
   $$\left(\frac{n_o}{n_i}\right)_1 = \frac{(V_o)}{(V_i)} = \frac{(\quad)}{(\quad)} = \underline{\quad}.$$

   Row 2 Data: Step-up transformer.
   $$\left(\frac{n_o}{n_i}\right)_2 = \frac{(V_o)}{(V_i)} = \frac{(\quad)}{(\quad)} = \underline{\quad}.$$

2. Find the current in each coil of Data Table 1. Use the equation: $I = V/R$ where $R_o = 100$ ohms, $R_i = 10$ ohms (or whatever resistance values you used), and V = voltage across each resistor.

Row 1 Data: Step-down transformer.

$$I_i = \frac{V_R\text{-in}}{R_i} = \frac{(\quad V)}{10\ \Omega} = \underline{\quad}\ A$$

$$I_o = \frac{V_o}{R_o} = \frac{(\quad V)}{100\ \Omega} = \underline{\quad}\ A$$

Row 2—Data: Step-up transformer.

$$I_i = \frac{V_R\text{-in}}{R_i} = \frac{(\quad V)}{10\ \Omega} = \underline{\quad}\ A$$

$$I_o = \frac{V_o}{R_o} = \frac{(\quad V)}{100\ \Omega} = \underline{\quad}\ A$$

3. Find the input and output power to the transformer assembly, for each row of Data Table 1. Then find the efficiency of the transformer.

Row 1 Data: Step-down transformer.

$$P_{in} = V_i \times I_i = \underline{\quad}\ W$$

$$P_{out} = V_o \times I_o = \underline{\quad}\ W$$

$$Eff = \frac{P_o}{P_i} \times 100\ \% = \underline{\quad}\ \%$$

$$Eff = \frac{(\quad)}{(\quad)} \times 100\ \% = \underline{\quad}\ \%$$

Row 2 Data: Step-up transformer.

$$P_{in} = V_i \times I_i = \underline{\quad}\ W$$

$$P_{out} = V_o \times I_o = \underline{\quad}\ W$$

$$Eff = \frac{P_o}{P_i} \times 100\ \% = \frac{(\quad)}{(\quad)} \times 100\% = \underline{\quad}\ \%$$

4. Find the power loss for each transformer assembly.

Row 1 Data: Step-down transformer.

$$P_i - P_o = P_{loss} = (\underline{\quad}) - (\underline{\quad}) = \underline{\quad}\ W$$

Row 2 Data: Step-up transformer.

$$P_i - P_o = P_{loss} = (\underline{\quad}) - (\underline{\quad}) = \underline{\quad}\ W$$

WRAP-UP

Use the data from Data Table 1, and the results of the *Calculations* you've just completed to complete Data Table 2. For the entry labeled "Turns Ratio" enter the ratio based on the vendor/teacher-supplied data, **not** that obtained in Step 1 of *Calculations*.

DATA TABLE 2

Input or Primary Coil	Output or Secondary Coil	Transformer Type (Step-up or step-down)	Turns Ratio n_o/n_i (Ideal)	Voltage Ratio V_o/V_i	Current Ratio I_o/I_i	Power			Eff %
						In	Out	Loss	
B	A								
A	B								

Conclusions

1. Compare the turns ratio n_o/n_i determined in Step 1 of Calculations with the true ratio for the transformer assembly (provided by your teacher) and listed in Data Table 2. Do the two ratios agree for either the step-up or step-down transformer arrangement? Explain—why or why not.

2. What is the **ideal** "electrical advantage" of the step-down transformer you used? Of the step-up transformer?

3. What is the **actual** "electrical advantage" of the step-down transformer? Of the step-up transformer?

4. How did the current ratio (I_o/I_i) compare with the voltage ratio (V_o/V_i) for each transformer?

Student Challenge ————————————————————————

1. You are to "drive" an AC motor that requires 15 volts and 1 ampere of current with the transformer you used. Would you use the step-down or step-up arrangement if 30 volts AC were available for input voltage to the transformer?

2. Suppose that the transformer is 85% efficient. For the same situation as described above, how much current would be needed on the input side to ensure that the motor did indeed receive 15 watts of power (that is, 15 volts × 1 amp)?

3. The resistors R_o and R_i are rated around 5 watts each. What precautions should you take when adjusting the AC power supply to ensure that you will not overload the resistors and possibly burn them out?

Review

1. *View and discuss the video, "Force Transformers in Electrical Systems."*

2. *Review the Objectives and the Main Ideas of the print materials in this subunit.*

3. *Your teacher may give you a test over Subunit 4, "Force Transformers in Electrical Systems."*

Summary

LEARNING PATH

1. **Read the Summary of Unit 7, "Force Transformers."**
2. **View and discuss the video, "Summary: Force Transformers."**
3. **Your teacher may give you a test over Unit 7, "Force Transformers."**

MAIN IDEAS

- *Force transformers change a given input into a desirable output.*
- *Force transformers usually amplify an input such as force, torque, displacement or speed. This amplification gives workers a mechanical advantage.*
- *Ideal transformers are those where there is no resistance. For ideal transformers, "Work In" equals "Work Out." Almost all transformers encounter resistance while operating. For actual transformers, "Work Out," is always less than "Work In." Therefore, efficiency is always less than 100 percent.*
- *In this unit, mechanical, fluid and electrical force transformers were studied. They were found to amplify force, torque, pressure and voltage.*

You've seen that "force" transformers are a class of machines or devices in mechanical, fluid and electrical systems that amplify an input to obtain a desired output.

The input that's amplified may be force, torque, pressure, voltage—or even displacement. In each application of a force transformer, something is always amplified or gained at the expense of something else.

For example, in a block and tackle, force is amplified. But the input force has to move through a longer distance than the load does at the other end. Force is amplified at the expense of distance.

In a belt-and-pulley system, torque may be amplified at the expense of angular speed. In a pressure booster, pressure may be amplified at the expense of fluid volume displaced. In a voltage transformer, if voltage is amplified, current is reduced. In other words, there's always a trade-off.

The schematic below sums up the common (unifying) action of all force transformers.

SOURCE (INPUT)	⇨	COUPLING DEVICE (TRANSFORMER)	⇨	LOAD (OUTPUT)

In this unit, you've seen several types of force transformers. Table 7-7 below lists the types, what's amplified, and one useful equation for ideal mechanical advantage.

TABLE 7-7: TYPES OF FORCE TRANSFORMERS

Type	What Is Amplified	Ideal Mechanical Advantage
Block and Tackle	Force	$IMA = \dfrac{D_i}{D_o}$
First-class Lever	Force	$IMA = \dfrac{L_i}{L_o}$
Inclined Plane	Force	$IMA = \dfrac{D_i}{D_o}$
Wedge	Force	$IMA = \dfrac{2h}{b}$
Wheel and Axle	Force	$IMA = \dfrac{r_i}{r_o}$
Belt Drive	Torque	$IMA = \dfrac{\omega_i}{\omega_o} = \dfrac{r_o}{r_i}$
Gear Drive	Torque	$IMA = \dfrac{N_o}{N_i}$
Hydraulic Jack	Force	$IMA = \dfrac{A_o}{A_i}$
Pressure Intensifier	Pressure	$IMA = \dfrac{A_i}{A_o}$
Voltage Transformer	Voltage	$IMA = IEA = \dfrac{n_o}{n_i}$

Note: D_i, D_o are distances moved at the input and output.
r_i, r_o are radii of wheels, axles, pulleys at input and output.
L_i, L_o are lever arms of input and output forces.
ω_i, ω_o are angular speeds at input and output.
N_i, N_o are gear teeth at input and output.
h, b are the height and width of a wedge.
A_i, A_o are the piston face areas at the input and output.
n_i, n_o are the number of coil windings at the input and output.

For each transformer, when friction or resistance is absent, "Work In equals Work Out" or "Power In equals Power Out." That makes the transformer IDEAL and the efficiency 100%. But for almost all force transformers, some friction—or resistance—is present. So the operation isn't 100% efficient. The equations on the next page sum up these ideas.

Ideal Transformers (no resistance)

Work In = Work Out

Power In = Power Out

Eff = 100%

Actual Transformers

Work Out Less Than Work In

Power Out Less Than Power In

$$Eff = \frac{AMA}{IMA} \quad \text{(always less than 1.00)}$$

or

$$Eff = \frac{AMA}{IMA} \times 100\% \quad \text{(always less than 100\%)}$$

How much have you learned about force transformers in mechanical, fluid and electrical systems?   Take another look at the devices and machines shown in Figures 7-1 and 7-6. (Both are in the "Overview" of this unit.) How many and what types of force transformers can you identify in each picture? Can you tell what is being amplified (or transformed) in each picture?

Glossary

Belt Drive: A rotational mechanical force transformer that has two wheels of different radii. The two wheels are mounted on different axles that are interconnected by a belt or chain.

Block and Tackle: An arrangement of pulleys that amplifies force.

Drive System: A class of rotational mechanical force transformers used to produce changes in torque and angular speed.

Electrical Advantage: The ratio of output turns to input turns for a step-up or step-down electrical transformer. The ideal electrical advantage equals the actual electrical advantage only when there are no resistance-related losses in the transformer.

Force Transformer: A class of machines or devices in mechanical, fluid and electrical energy systems that change input values of force, movement or rate into different output values of the same quantities.

Gear Drive: A rotational mechanical force transformer that transforms torques and speeds by meshing toothed wheels.

Grade: The ratio of the rise of an incline to its length.

Ideal Efficiency: The efficiency expected in a force transformer when no resistance is present: 100 percent.

Idler Gear: Any intermediate gear in a gear train whose overall effect is to transfer motion without changing mechanical advantage.

Inclined Plane: A linear mechanical force transformer that enables one to apply a smaller force over a longer distance along an incline when moving an object from one elevation to another.

Incompressible Fluid: A liquid that can't be forced into a smaller volume.

Isolation Transformer: An electrical force transformer that has a mechanical advantage of one, such that voltage in equals voltage out, used to block DC signals from reading the secondary windings.

Lever: A simple machine that amplifies force. Depending on the relative locations of the input force, output force and fulcrum, it can be classified as a first-class, second-class or third-class lever.

Mechanical Advantage: The ratio of the output/input values of a certain parameter —such as force, torque speed, etc.—for a force transformer. When resistance is ignored, the mechanical advantage is referred to as "ideal." When the effects of resistance are included, the mechanical advantage is referred to as "actual."

Pressure Intensifier: A fluid force transformer that boosts or amplifies pressure from input to output.

Pulley: A linear mechanical force transformer that makes use of a rope and wheel.

Ratio: A means of expressing the relative magnitudes of two values in the forms "value 1/value 2" or "value 1:value 2."

Step-down Transformer: An electrical force transformer that has an output voltage less than the input voltage.

Step-up Transformer: An electrical force transformer that increases the output voltage level over the input voltage level.

Wheel and Axle: A rotational mechanical force transformer that consists of two wheels of different radii mounted on a common shaft.

GENERAL REVIEW: UNITS 5 THROUGH 7

Unifying Concepts
For Energy, Power, and Force
Transformers in Four Energy Systems

GENERAL REVIEW OBJECTIVES

When you have finished this General Review of Units 5-7—"Energy," "Power" and "Force Transformers"—you should be able to do the following:

1. *Describe energy, power and force transformers as they apply in the energy systems you've studied.*

2. *Give a unifying statement (or statements) for energy, power and force transformers that helps you understand how they apply in different energy systems.*

3. *Use the unifying statements to write an equation—or formula—that applies to each energy system.*

LEARNING PATH

1. *Read the "General Review" section.*

2. *View and discuss the video overviews for the General Review of Units 5-7 of <u>Principles of Technology</u>.*

3. *Participate in class discussions.*

MAIN IDEAS

- *A single unifying statement of a concept or principle (such as energy, power, or force transformers) can help you understand how each of these concepts is applied in the various energy systems.*

- *A single idea or expression that can be applied effectively to several energy systems broadens our knowledge of the basic meaning of a concept or principle.*

Principles of Technology organizes basic principles found in many technologies. These principles are found in mechanical, fluid, electrical and thermal energy systems.

The *Principles of Technology* units you've studied cover the concepts of force, work, rate, resistance, energy, power, and force transformers. They're the first seven of 14 units in this course. Each concept is identified with a unifying statement—or unifying equation—that's valid in several energy systems.

At the end of Unit 4, *Resistance*, you reviewed the first four concepts of force, work, rate and resistance. Let's go back and examine the unifying statement for each one briefly. (A more detailed examination can be found by reviewing the material provided in the "General Review" at the end of Unit 4.)

Unit 1, *Force*

   The unifying statement can be written as:

   Force and forcelike quantities tend to cause displacement—or movement— of something (mass, fluid, charge or heat energy).

   This statement may be applied to all four energy systems.

Unit 2, *Work*

   The unifying statement can be written as:

   Work results when a forcelike quantity causes a "displacement," and is equal to the product of the forcelike quantity times the "displacement."

   This statement applies to all of the energy systems except thermal systems.

Unit 3, *Rate*

   The unifying statement can be written as:

   Rate is the ratio of the change in a displacementlike quantity to the time required for that change to occur.

   This statement applies to all four energy systems.

Unit 4, *Resistance*

   The unifying statement can be written as:

   Resistance opposes motion. Resistance causes motion to eventually cease if no forcelike quantities are present. Resistance equals a force or forcelike quantity divided by a rate.

   This statement applies in all four energy systems.

Now let's sum up Units 5-7, *Energy*, *Power*, and *Force Transformers*. We'll establish the unifying statement for each concept. At the same time, we'll discuss how useful relationships within each concept lead to equations that give quantitative answers to problems you can solve.

Unit 5, *Energy*

   The total energy in a system can be accounted for at all times. Energy in machines often is available in the form of potential or kinetic energy. Total energy equals the sum of potential and kinetic energy plus any losses, generally heat.

Many variations of potential and kinetic energy exist. These variations depend on the device or machine involved. However, two important equations that express content of energy are:

   Gravitational Potential Energy = Weight × Height Above a Reference Position

   Kinetic Energy = $\frac{1}{2}$ × Mass × Speed Squared

Table 7-8 sums up energy as a unifying concept in the four energy systems.

TABLE 7-8: ENERGY AS A UNIFYING CONCEPT

Energy System	Potential Energy (E_p)	Kinetic Energy (E_k)
MECHANICAL Translational Rotational	Gravitational $\quad E_p = mgh$ or $E_p = wh$ Elastic $\quad E_p = \frac{1}{2} kd^2$ ------	$E_k = \frac{1}{2} mv^2$ $E_k = \frac{1}{2} I\omega^2$
FLUID	$E_p = (\rho V) gh$ where: $m = \rho V$	$E_k = \frac{1}{2} (\rho V) v^2$
ELECTRICAL	$E_p = \frac{1}{2} C (\Delta V)^2$ $E_p = \frac{1}{2} LI^2$	------ ------
THERMAL	When there's resistance in an operating machine or system, some useful energy (in the form or potential or kinetic energy) must overcome resistance. The lost energy usually shows up as heat—or thermal energy.	

Note: In fluid systems, V represents a volume. In electrical systems, V represents a voltage.

Unit 6, *Power:*

Power is a special kind of rate. Power also is an important concept in technology. The two unifying formulas for power in mechanical, fluid and electrical systems are:

$$\text{Power} = \text{Work divided by Time}$$

$$\text{Power} = \text{Forcelike Quantity} \times \text{Rate}$$

From these two relationships, you get useful equations for power in mechanical, fluid and electrical systems. Tables 7-9 and 7-10 sum up these equations.

TABLE 7-9. UNIFYING FORMULA: POWER = WORK/TIME

Energy System	Forcelike Quantity	Work	Power Equation
MECHANICAL Linear	 Force (F)	 $W = F \times D$	 $P_M = \dfrac{F \times D}{t}$
Rotational	Torque (T)	$W = T \times \theta$	$P_M = \dfrac{T \times \theta}{t}$
FLUID	Pressure Difference (Δp) or Pressure (p)	$W = (\Delta p) \times V$ $W = p \times (\Delta V)$	$P_F = \dfrac{(\Delta p) \times V}{t}$ $P_F = \dfrac{p \times (\Delta V)}{t}$
ELECTRICAL	Voltage Difference (ΔV)	$W = (\Delta V) \times q$	$P_E = \dfrac{(\Delta V) \times q}{t}$

Note: In fluid systems, V represents a volume. In electrical systems, V represents a voltage.

TABLE 7-10. UNIFYING FORMULA: POWER = "FORCE" × RATE

Energy System	Forcelike Quantity	Rate	Power Equation
MECHANICAL Translational Rotational	Force (F) Torque (T)	Speed (v) Angular Speed (ω)	$P_M = F \times v$ $P_M = T \times \omega$
FLUID	Pressure Difference (Δp) or Pressure (p)	$Q_V = \dfrac{V}{t}$	$P_F = (\Delta p) \times Q_V$ $P_F = p \times Q_V$
ELECTRICAL	Voltage Difference (ΔV)	$I = \dfrac{q}{t}$	$P_E = (\Delta V) \times I$

Note: In fluid systems, V represents a volume. In electrical systems, V represents a voltage.

Thermal power is defined just like thermal rate. Thermal power also is measured in the same units as thermal rate

$$\text{Thermal Power} = \frac{\text{Heat Energy Produced}}{\text{Time}}$$

or

$$P = \frac{H}{t}, \text{ in cal/sec or Btu/sec}$$

Unit 7, *Force Transformers*

Force transformers are machines. These machines vary in form and have many uses. During your study of Unit 7, *Force Transformers*, you found that the unifying concept for a force transformer helps you describe what happens when a machine is used to gain a mechanical advantage.

You found that the basic idea of "Input → Transformer → Load" can be used to express the unified concept of a force transformer in mechanical, fluid and electrical systems. That basic idea is shown below.

Ideal transformers are 100% efficient. However, in the real world of machines, there's always some resistance present. So efficiency is less than 100%. These main ideas are summed up as follows:

Ideal Transformers (no resistance)

Work In = Work Out

Power In = Power Out

Efficiency = 100%

Actual Transformers

Work Out Less Than Work In

Power Out Less Than Power In

$$\text{Eff} = \frac{\text{AMA}}{\text{IMA}} \times 100\% \text{ (always less than 1.00\%)}$$

Finally, force transformers usually amplify something to provide a mechanical advantage. The types of force transformers studied and what they amplify—with a representative equation for **ideal** mechanical advantage—are summarized as follows:

TABLE 7-11: TYPES OF FORCE TRANSFORMERS

Type	What Is Amplified	Ideal Mechanical Advantage
Block and Tackle	Force	$\text{IMA} = \dfrac{D_i}{D_o}$
First-class Lever	Force	$\text{IMA} = \dfrac{L_i}{L_o}$
Inclined Plane	Force	$\text{IMA} = \dfrac{D_i}{D_o}$
Wedge	Force	$\text{IMA} = \dfrac{2h}{b}$
Wheel and Axle	Force	$\text{IMA} = \dfrac{r_i}{r_o}$
Belt Drive	Torque	$\text{IMA} = \dfrac{\omega_i}{\omega_o} = \dfrac{r_o}{r_i}$
Gear Drive	Torque	$\text{IMA} = \dfrac{N_o}{N_i}$
Hydraulic Jack	Force	$\text{IMA} = \dfrac{A_o}{A_i}$
Pressure Intensifier	Pressure	$\text{IMA} = \dfrac{A_i}{A_o}$
Voltage Transformer	Voltage	$\text{IMA} = \text{IEA} = \dfrac{n_o}{n_i}$

Note: D_i, D_o are distances moved at the input and output.
r_i, r_o are radii of wheels, axles, pulleys at input and output.
L_i, L_o are lever arms of input and output forces.
ω_i, ω_o are angular speeds at input and output.
N_i, N_o are gear teeth at input and output.
h, b are the height and width of a wedge.
A_i, A_o are the piston face areas at the input and output.
n_i, n_o are the number of coil windings at the input and output.

End-of-unit Student Exercises
Force Transformers

1. "Force transformers" basically work the same in mechanical, fluid and electrical systems. Regardless of the energy system, the action always involves a source or input, a coupling device and ____.

2. Force transformers are a class of machines and devices in mechanical, fluid and electrical energy systems that change ____ (input, output) values of "force," movement or rate into different ____ (input, output) values.

3. When output work equals input work in a force transformer, the force transformer ____ (has lost energy through friction, is 100% efficient).

4. Each of the three drawings below shows an effort, a load, and a pivot for a lever. Complete the sentence below each drawing to correctly identify the class of lever being shown.

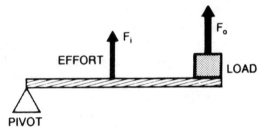

a. An arm is a ____-class lever.

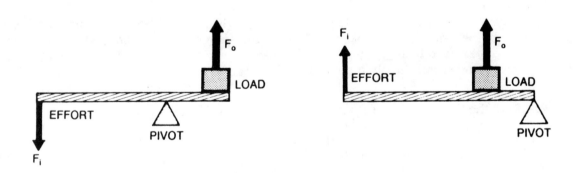

b. A crowbar is a ____-class lever. c. A wheelbarrow is a ____-class lever.

5. A mechanical force transformer has an efficiency equal to the ratio of the ____ (Work Out/Work In, Work In/Work Out) times 100%.

6. When 200 ft·lb of input work is done on a *frictionless* block and tackle, the lifting force on the load raises the load 6 inches (0.5 foot). What's the amount of lifting force on the load?

7. From a count of the number of cables that support a load, the ideal mechanical advantage (IMA) of a block-and-tackle force transformer is found to be equal to 2. When the 40,000-lb load is actually lifted, it requires 22,000 lb of input force. The efficiency of the block and tackle is about ____.

   (**Hint:** Remember that Eff $= \dfrac{\text{AMA}}{\text{IMA}} \times 100\%$, where AMA $= F_o/F_i$.)

8. Match the "device" in the left column with the general type of force transformer described in the right column.

   a. belt-and-pulley system
   b. autotransformer
   c. crowbar
   d. power-brake booster

   (1) linear force transformer
   (2) fluid pressure transformer
   (3) rotational torque transformer
   (4) electrical voltage transformer

9. A chain hoist with an 8-inch-diameter input wheel and a 2-inch-diameter output or load wheel operates without internal friction (100% efficient). Find the actual mechanical advantage of the chain hoist. (IMA and AMA for wheel-and-axle force transformers are: IMA $= \dfrac{\text{input radius}}{\text{output radius}}$ and AMA $= \dfrac{\text{output force}}{\text{input force}}$.)

10. An overhead-crane hoist was designed with an IMA of 10 when rigged as specified. Internal friction reduces the efficiency to 80%. Find the load that the crane will lift when the input force on the line is 15 tons. (See "Hint" in Problem 7.)

11. A belt-drive system has an ideal mechanical advantage of 3. If the smaller wheel of the belt drive is on the input shaft, the larger output wheel will cause the output shaft to turn ____ (faster, slower) and the output shaft will have ____ (more, less) torque than the input shaft.

12. Find the IMA of a simple gear train made up of the following 3 gears.
   Gear 1 (input) has 12 teeth
   Gear 2 (middle) has 42 teeth
   Gear 3 (output) has 30 teeth

   IMA $= \dfrac{N_o}{N_i}$ where: N $=$ number of teeth

13. In a single-acting hydraulic pressure booster with internal resistance, the ratio of pressure output divided by pressure input is (308 psi/100 psi). The IMA for this booster is 4. Find its efficiency. (See "Hint" in Problem 7.)

14. For the hydraulic jack shown, mark those quantities that (1) increase, (2) remain the same, or (3) decrease, from *input to output*. (See drawing for reference.)

   a. pressure   (1) increases
   b. force   (2) remains
   c. piston movement   the same
   d. piston area   (3) decreases
   e. volume displaced

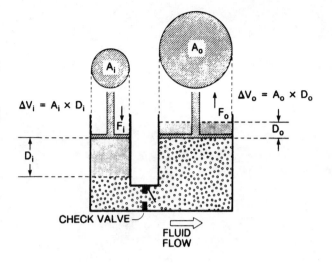

15. For the pressure intensifier shown here, indicate those quantities that (1) increase, (2) remain the same, or (3) decrease, from *input to output*.

   a. pressure   (1) increases
   b. force   (2) remains
   c. piston movement   the same
   d. piston area   (3) decreases
   e. volume displaced

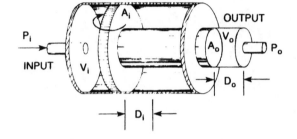

16. Electrical transformers can be used to step up or step down ____ (voltage, current, voltage or current).

17. Find the voltage output of an electrical voltage transformer with 500 turns on the primary or input winding, and 900 turns on the secondary, or output winding, when 8 volts AC is applied to the input winding. (Assume the transformer is 100% efficient, IEA = n_o/n_i and IEA = V_o/V_i.)

18. Find the current output for the transformer in Problem 17 when the AC current input is $I_i = 0.4$ A.